Cities in the Third Wave

Cities in the Third Wave

The Technological Transformation of Urban America

Second Edition

Leonard I. Ruchelman

ROWMAN & LITTLEFIELD PUBLISHERS, INC.
Lanham • Boulder • New York • Toronto • Plymouth, UK

ROWMAN & LITTLEFIELD PUBLISHERS, INC.

Published in the United States of America
by Rowman & Littlefield Publishers, Inc.
A wholly owned subsidiary of The Rowman & Littlefield Publishing Group, Inc.
4501 Forbes Boulevard, Suite 200, Lanham, Maryland 20706
www.rowmanlittlefield.com

Estover Road, Plymouth PL6 7PY, United Kingdom

British Library Cataloguing in Publication Information Available

Library of Congress Cataloging-in-Publication Data
Ruchelman, Leonard I., 1933–
  Cities in the third wave : the technological transformation of urban America /
Leonard I. Ruchelman. — 2nd ed.
    p. cm.
  Includes bibliographical references and index.
  ISBN-13: 978-0-7425-3908-2 (cloth : alk. paper)
  ISBN-10: 0-7425-3908-3 (cloth : alk. paper)
  ISBN-13: 978-0-7425-3909-9 (pbk. : alk. paper)
  ISBN-10: 0-7425-3909-1 (pbk. : alk. paper)
    1. Cities and towns—Effect of technological innovations on—United States.
  2. Information technology—Social aspects—United States. 3.
  Telecommunication—Social aspects—United States. I. Title.
  HT167.R83 2007
  307.760973—dc22

                                                                    2006026699

Printed in the United States of America

♾™ The paper used in this publication meets the minimum requirements of
American National Standard for Information Sciences—Permanence of Paper
for Printed Library Materials, ANSI/NISO Z39.48-1992.

To my wife, my children, and my grandchildren

Contents

Figures, Tables, and Boxes

FIGURES

TABLES

BOXES

Preface

In the belief that technology is a force that has created and recast cities throughout history, *Cities in the Third Wave, 2nd Ed.*, attempts to address the important questions of how cities are presently being affected by new technology and what they will be like in the future. Countries such as the United States and Japan, among select others, have passed through the preindustrial and industrial stages of urban development and have now entered postindustrialism—what futurists call the "third wave."

This book surveys the remarkable transformation that is taking place in urban America and asks: How do computers and communications technologies that are fueling an information economy affect cities and suburbs? How do urban places adapt to changing conditions brought about by deindustrialization and the globalization of business enterprise? What kinds of strategies do they devise to attract and retain investment and jobs? Why do some cities appear to prosper in this new postindustrial era while others become victims or mere survivors? This book will help readers understand what it will take for their cities, and other cities, to survive and even thrive in this changing environment.

I gratefully acknowledge the advice and assistance provided by many of my colleagues at Old Dominion University, which is located in the developing metropolitan area of Hampton Roads, Virginia. Of immense help, also, were the discussions and brainstorming that took place in my graduate classes, which were sources of inspiration for me to undertake this project in the first place.

1

Technology and Urban Development

Technology does not stand alone in stimulating urbanization, but it is one of the more critical factors in accounting for the creation and proliferation of cities in the United States and much of the Western world. Today, North American cities are in the midst of a technological revolution. Driven largely by advances in transportation and communications, the nation's urban areas are being profoundly reshaped, portending dramatic changes in the way most Americans will live and work. But although there has been much speculation about the impact of emerging technological change on society in general, little is known about its potential effects on urban conditions.

As the new millennium takes firm hold, old ideas about the development and management of the contemporary city seem less and less useful. Conventional wisdom about the nature of space, time, and distance as they affect human settlement are being questioned as never before. Physical boundaries seem less relevant, and traditional distinctions between private and public are similarly being challenged. Consequently, urban life seems more volatile and bewildering than at anytime in recent history. Our goal in this book is to offer a balanced perspective on these seismic shifts by charting the technological transition of older agricultural communities, first to manufacturing-dominated cities and then to the more recent phenomenon of cities dominated by information and communications.

THEORETICAL PERSPECTIVES ON
THE ROLE OF TECHNOLOGY

Technology is typically part of a highly complex process involving myriad decisions by individuals and organizations on matters of location, resources, and demographics. Decisions on how cities should be formed, however, are influenced to a significant extent by the technological possibilities of the time. At the turn of the century and right to the present, a number of theorists and writers have emerged propagating ideas on how technological innovation could best be managed for guiding the development of cities. Many of these ideas have served as the basis of city planning and continue to influence development decisions in both the private and public sectors.

Most prominent among the early writers is an Englishman by the name of Ebenezer Howard, whose work continues to shape the field of urban planning. Although his occupation was court reporter, he is considered one of the founders of what the English call "town and country planning." Howard observed how poor people lived in the industrial cities of late-nineteenth-century England and did not like what he saw. Noise, dirt, and disease made for human debasement and unhealthy living. As he understood it, much of the problem could be attributed to crowded living conditions in city slums. His vision for saving the people was to bring the industrial city to an end.

In 1898, Howard published his book *Tomorrow: A Peaceful Path to Real Reform*, revised in 1902 as *Garden Cities of Tomorrow*. His goal was to halt the growth of large industrial cities and repopulate the countryside by building a new kind of community, which he called the Garden City. He envisaged his Garden City as a tightly organized urban center housing approximately thirty thousand inhabitants and surrounded by a perpetual "green belt" of farms and parks. Within the city there would be both quiet residential neighborhoods and, in the center, facilities for a full range of commercial, industrial, and cultural activities.

Howard did not view the Garden City as a "satellite town" perpetually serving the larger central city. Rather, he believed that large metropolitan centers would eventually shrink to insignificance as their inhabitants sought a new way of life in a decentralized society. Ultimately, the urban population would be distributed among hundreds of Garden Cities, whose small scale and diversity of functions would embody a society in which the ordinary man would finally win out. Railways and canals linking the Garden Cities to each other and to the larger central city would promote a certain amount of commerce and trade.

The "agricultural belt" was expected to play an important role in the economy of the Garden City. Farmers who lived there would supply the town with most of its food. Because transportation costs would be almost nonexistent, the farmer would receive a good price and the consumer would get fresh vegetables and dairy products at reduced cost. The agricultural belt, moreover, would prevent the town from sprawling out into the countryside and would ensure that its citizens enjoyed both a compact urban center and an ample, open countryside.

The problem of unhealthy cities was especially important to Howard. As a health measure, he located the factories at the periphery of the city adjacent to the circular railroad surrounding the town and away from residential areas. Here would be located enterprises most suitable to a decentralized society: the small machine shop, the cooperative printing works, the jam factory where fruits from the rural cooperatives were to be processed.

Howard's utopian vision of the Garden City was and still is the basis of an immensely influential planning concept. Howard succeeded in getting two garden cities, Letchworth and Welwyn, built in England. Since World War II, Sweden and other Scandinavian countries have built a number of satellite towns modeled on Garden City principles. In the United States in the 1920s, Howard's ideas were promoted by a group of dedicated followers, among them Lewis Mumford, Clarence Stein, and Catherine Bauer. Examples of successful adoption, though often in modified form, were Radburn, New Jersey; Greenbelt, Maryland; and Forest Hills, New York. More recently, in the 1960s, Reston, Virginia, and Columbia, Maryland, were designed and built as planned communities by developer James W. Rouse.

While Howard was attempting to neutralize the negative effects of the industrial city, others who wrote at the turn of the century viewed technology as having a more pervasive impact on cities. As early as 1902, H. G. Wells wrote that giant cities in Western society were in all probability destined to such a process of dissection and diffusion as to amount almost to obliteration within a measurable further space of years. According to Wells, the city will diffuse itself until it has taken up considerable areas and many of the characteristics of what is now country, while the country will take upon itself many of the qualities of the city. The old antithesis will cease, and the boundary lines will altogether disappear.[1]

In the late 1920s, Frank Lloyd Wright elaborated on this theme in his plan for a new city. The stock market crash of 1929 and the Great Depression reinforced his belief that the country needed a radical change in its physical and economic organization. The complete plan was published in 1932 under the title *The Disappearing City*. It describes a time when the great urban centers are not merely disappearing but have already ceased to exist. Wright believed that this was inevitable in the age of the automobile and the telephone. Great concentrations of people in cities were wasteful and unnecessarily burdensome when modern means of communication and transportation could overcome distance.

Like so many Americans of his day, Wright was fascinated by the automobile and convinced of its potential to revolutionize modern life, for the automobile created all kinds of dizzying possibilities based on a new mastery of time and space. He believed that deconcentration, if taken to its logical end, could create the material conditions for a nation of independent farmers and proprietors. If properly planned, cities could spread far over the landscape and still not lose their cohesion or efficiency. The diffusion of population would create conditions for the universal ownership of land. A society where wealth and power were highly concentrated in the hands of the few would be transformed; the means of production would be held by the many. People would be able to function as part-time farmers, part-time artisans, part-time mechanics, and part-time intellectuals. Thus, technology would point the way for a revival of the democratic goals of an earlier agrarian society. He envisioned a Broadacre City that would be seen as a juxtaposition of the past and the present: the ideals of eighteenth-century Jeffersonian democracy given new meaning through twentieth-century technology.

While Ebenezer Howard wrote of the "marriage of town and country," he made a clear distinction between the two realms. Frank Lloyd Wright, in contrast, believed there must be no distinction between rural and urban living. This is to say, the physical separation between rural and urban areas would be outmoded. In Broadacre City, homes, factories, stores, office buildings, and cultural centers would be distributed throughout farmland and forests. All of this would be made possible by a transportation system that would allow citizens to engage in business and social interchange conveniently and expeditiously. The blending of town and country, physical and intellectual labor, work and leisure was all part of Wright's vision for eliminating the fragmentation of modern life.

Many of Howard's and Wright's ideas were adapted in the United States and served as prototypes for new town planning and later suburban development. Together with Lewis Mumford, Clarence Stein, and Catherine Bauer, they came to be called the Decentrists because of their view that great cities should be decentralized and thinned out. Mumford's *The Culture of Cities* (1938) was a diatribe against large urban centers.[2] His critique of the metropolis focused on the quality of life available to the urban citizen. Of special concern was the work environment, which centered around the machine and in which workers either served as babysitters to the machine or operated in a machine-like social organization. Highways transported people in sealed vehicles at high speeds through the central city to the contrived social settings of the suburbs.

In Mumford's view, the industrial city is deficient because it is inimical to human needs. Tall buildings, elevated highways, and industrial pollution all pose special hazards to human well-being. The tragedy of the modern metropolis, he argued, is that it "has become self-destructive of human values: everywhere the machine holds the center and the personality has been pushed to the periphery."[3] Later in his life, he became one of the persistent critics of overbuilding in New York City, presciently contending that it would result in a "sprawling mass of expressways, cloverleaves, bridges, viaducts, airports, garages, operating in an urban wasteland."[4]

To overcome these destructive features, Mumford advocated the planned organic community. In his view, a city was more than a collection of buildings; it was a living entity. As an architect, he insisted on reviewing buildings in terms of their effects on the larger whole as much as for their aesthetics. He believed that the more a city was planned in a low-rise, medium-density way, the better the life within it would be. On the neighborhood level, he advocated a variety of facilities enabling people in various phases of life to satisfy divergent needs: local hospitals for infants and the aged, space for adolescents and adults to recreate, and local shops to satisfy consumer needs. The location of these facilities within the neighborhood, Mumford claimed, would not exclude opportunities for participation in metropolitan affairs such as mass entertainment and symphonies, but would allow individuals near each other to satisfy personal needs as well as to assist others in meeting theirs.

In counterbalance to the Decentrists, in the 1920s the European architect Le Corbusier devised his own version of the industrial city that he called the Contemporary City.[5] The Contemporary City was composed not of the low buildings proposed by the Decentrists but of skyscrapers situated in a parklike setting. The center of the city was to function as a multilevel transportation complex that could facilitate the rapid movement of people through space. Two great superhighways crossed there, and below them was the station where subway lines intersected. Above the highways, mounted on great steel pillars, stood the main railroad terminal, and on the roof of this huge structure stretched a runway where airplanes could land.

In a more profound sense, Le Corbusier believed that the city existed for interchange: the most rapid possible exchange of ideas, information, and talents. Only the dense concentration of a large metropolis could assure a critical mass of creative juxtapositions, and this was the special function of urban life. The central terminal thus became the symbol of the modern city, where everything was in motion and where speed was the only constant.

In his Contemporary City, twenty-four glass-and-steel skyscrapers, each sixty stories high, surrounded the central terminal. They housed the business elite and the bureaucracies that provided managerial leadership to the city and to whole nations. In this scenario, there were no "corridor streets," as he called them, no more narrow roadways filled with traffic, tightly lined with five- or ten-story buildings. Instead, the "streets" were elevators, rising straight up instead of spreading out over a whole district. Although 500,000 to 800,000 people could work in the twenty-four sky-scrapers, the buildings covered less than 15 percent of the ground in the business center. The rest contained parks and gardens. The skyscrapers thereby freed the ground for open space and greenery. As conceived by Le Corbusier, the park was not in the city, the city was in the park. In this way, he was able to reconcile what at first glance seemed to be opposites of urban design: density and open space.

Various features of Le Corbusier's utopian city have gradually been incorporated in numerous projects ranging from low-income public housing to office buildings. Cities like New York, Chicago, and San Francisco have ever since been engaged in a competition to see who can build higher. Only belatedly have public housing agencies come to recognize that high-rise public housing does not work very well for low-income families, and many such projects are now in the process of being abandoned.

Perhaps the severest critic of all these schools is Jane Jacobs.[6] When tested in the real world, she argues, none of their plans for cities work. High-rise low-income public housing projects have become centers of crime and hopelessness and are in many ways worse than the slums they were intended to replace. Middle-income projects, she claims, are marvels of dullness and regimentation. Similarly, luxury housing projects manifest a vapid vulgarity. Civic centers with promenades that go from no place to nowhere are often void of promenaders. Expressways create social barriers as they cut through neighborhoods. "This," she states, "is not the rebuilding of cities. This is the sacking of cities."[7]

To a certain extent, Jacobs's vision for the modern city is a throwback to cities in the preindustrial era, cities that were compact and offered a rich mix of diverse uses. She contends that to save the city, it is necessary to plan for vitality by stimulating the greatest range of diversity of uses throughout each neighborhood district. Jacobs specifies four basic conditions to achieve this:

1. A neighborhood district should serve more than one primary function and preferably more than two. This will generate a more dynamic environment by encouraging the presence of people on the street at different times of the day and night.
2. Because long city blocks tend to be socially isolating, they should be made short to better facilitate encounters among the users of a city neighborhood.

3. Neighborhoods should consist of a mix of buildings that vary in age and condition, including a good proportion of old ones. Because older buildings are less costly, they allow creative experimentation with new ideas.
4. Finally, in contrast to Ebenezer Howard and Lewis Mumford, Jacobs argues that the concentration of people is necessary to assure a critical mass of activity on the street.

Other apostles of urban technology are R. Buckminster Fuller and Daniel Burnham. Fuller argued that the resources made available by the universe—energy, materials, space—were finite and had to be seen in terms of their relationship to one another. With that in mind, he designed geodesic domes made up of tetrahedrons—that is, pyramidlike shapes comprising three triangles put together into domes or spheres. The geodesic dome was exhibited at Expo '67 in Montreal where a still-standing, two-hundred-foot-high version served as the United States Pavilion. Though Fuller's intent was to demonstrate how maximum space and energy efficiency could be created with a minimum of material, his ideas did not allow for individual variation and, consequently, were never adopted by American society.

It was in Chicago that the ideas of Daniel Burnham captured the interest of the city's business leaders who were planning the World's Fair in 1893. Labeled the Columbian Exposition, the fair introduced a Renaissance style of building, arraying one grandiose monument after another in the exposition park. Although critics called it a "White City of wedding-cake buildings," it captured the imagination of both planners and the public. It stimulated a movement called the City Beautiful, and Burnham became the leading City Beautiful planner. In the Chicago Plan of 1909, Burnham's idea was to create romantic parks and attractive waterfront landscapes along Lake Michigan and to enhance the city with huge plazas and broad thoroughfares. Clearly, Burnham believed the city was a place to escape from.[8] Though the architecture of the City Beautiful movement went out of style, it reemerged in the post–World War II years in the form of grandiose cultural centers such as the immense Lincoln Square project in New York City.

More recently, beginning in the 1960s, a number of theorists have come to recognize new forms of social, spatial, and economic organization that are transforming the city. Noting the effects of innovations in communications technology, some call this transformation Postindustrial, while others call it the Third Wave, the Global Village, or the World Information Economy. One of the very early writers to anticipate and analyze the effects of communications media on people and on society was Marshall McLuhan.[9] This Canadian educator wrote about mass media, and his observations on the human consequences of technological change in communications attracted widespread attention, making him one of the most celebrated and controversial commentators on popular culture.

For McLuhan, a communications medium determines ways of sensing and organizing experience through a particular mixture of the senses and activities. His basic thesis is that electronic technology is an extension of man's central nervous system, which restores the human family to a state of togetherness that he terms the "global village."[10] Phonetic writing, he explains, transformed the oral into the visual. Later, printing imposed a visual and private, individualized consciousness whose "high definition" encourages detachment and relative isolation in the reader. Then came television, which according to McLuhan is a "cool" medium because its indistinct image requires the viewer to "fill in the blanks" in a way that involves all the senses. Thus, media such as print or television have effects that outweigh the overt messages they transmit. Each medium of communication engenders a distinctive way of looking at the world; as he explains it, "The medium is the message." Television viewing is not a single-minded, linear process, as print reading is. Rather, it stimulates a mode of perception that affects many other aspects of life. McLuhan attributed to the influence of television such trends as the desire for small cars and for changes in church liturgy. However, critics assert that many of his predictions seem forced and that some have not been borne out—for example, his prediction that baseball would decline in popularity because it is a "linear" game.

Though McLuhan's theory of the role of media has not been validated with hard evidence, some observers find it highly suggestive for explaining changing perceptions in urban society. Joshua Meyrowitz, for instance, argues that the evolution of media and new information technologies has decreased the significance of the linkage between physical place and social "place."[11] Physical place defines a distinct situation because its known boundaries limit people's perceptions and interactions. Consequently, social situations vary, and people behave differently in each situation. New patterns of information flow, however, lead to the disassociation of social situations from their boundaries, thereby stimulating changes in the "sensory balance" of people. This, in turn, alters their behavior.

Everett M. Rogers explains that new information technologies accelerate the adoption of innovations and the interpretation of their uses and meaning.[12] New social activities that could influence urban form emerge as a result. To provide a hypothetical illustration, the increase of interactive communication in society is likely to generate new leisure activities, travel, and tourism, which then affect the function and use of urban space.

Though still in their infancy, other theories focus more directly on how new communications technology may influence urban form. Analyzing the transition from an industrial to a postindustrial economy, Alvin Toffler identifies a new mode of production he calls the "electronic cottage."[13] He speculates that the nation's biggest factories and office towers may someday stand half empty as production returns to the home by means of new, higher electronic technology. Toffler reasons that a major inducement to promote the electronic cottage is the economic trade-off between transportation and telecommunications. U.S. metropolitan areas are already experiencing major transportation problems, with mass transit systems severely strained and roads and highways heavily congested—and becoming even more so. While the costs of commuting are borne by individual workers, they are indirectly passed on to employers in the form of high wage costs and to consumers as higher prices. In addition, as the ratio of commuting time to working time continues to rise, we can expect growing resistance to commuting. Another payoff, according to Toffler, is that the electronic cottage will strengthen the nuclear family by putting working parents back into the home.

Building on the concepts of McLuhan and Toffler, Tarik A. Fathy contends that the revolution in the use and application of information will produce a "telecity."[14] The telecity model is based on interactive communication networks that can accommodate layers of "teleactivities"—interactive, individualized communications connecting persons, tasks, and information regardless of their actual locations. Online networks of information services, teleworking at home using microcomputers, and psychological neighborhoods (defined as mental landscapes that connect a virtual structure of activities) are emerging types of teleactivities. In light of this, Fathy foresees a shift in the image of the city in that many activities occurring in different places will seem to be located in proximity. Thus, a telecity will be "superimposed" over the physical city (or cities), making it difficult to perceive activities in their spatial structure from within the city.

The telecity concept, it should be noted, differs from an earlier concept of the "wired city." Wired cities were conceived as experimental projects that utilize two-way cable television along with other new information technologies.[15] A wired city can thus be understood as a small-scale community, artificially implemented. At last count, ten communities in more than five countries including the United States have developed these systems.

Taking a somewhat different tack, other theorists address the evolution of the "information society," or what Toffler refers to as the Third Wave.[16] In their writings, Daniel Bell and Alain Touraine observe that the information society is replacing the industrial society in developed countries.[17] The prime characteristics of the information society, according to Bell, are the transition from a goods-producing to a service economy and the increased emphasis on theoretical knowledge as the guiding force for social change. This is facilitated by a new class of professionals, engineers, technicians, and scientists that is coming to dominate economic decision making. Touraine describes a "programmed society," characterized by a new trend of business based on knowledge and organization, along with the growth of new bureaucratic and professional classes. Building on these earlier works, John Naisbitt examines the many ways in which the nation is restructuring.[18] Faster flows of information through all forms of interactive communication encompass several changes in the character of work, the growth of the knowledge class, movement from a national to a global economy, technological change in the direction of high-tech industry, and replacement of centralized structures with decentralized ones. All these factors have implications for how cities of the future will develop.

An important work in this regard, one that attempts to provide a blueprint of the future direction of postindustrial urban society, is Saskia Sassen's *The Global City: New York, London, Tokyo.*[19] According to Sassen, advancements in telecommunications are transforming modern capitalism into a global network of both corporations and cities. At the top of the global urban hierarchy are first-tier global cities such as New York, London, and Tokyo. These cities have extensive concentrations of high-level corporate decision makers representing institutions of finance, industry, commerce, law, and media at the global level. As such, global cities constitute a system primarily in terms of international finance, investment, and real estate markets. In this capacity, they serve as the brains of the global economy. Sassen observes, however, that only a handful of cities are in the first tier. Though most cities occupy niches lower down in the world urban hierarchy and perform diverse economic functions, the greater number are interconnected through a web of transnational corporations and their suppliers. Sassen also notes that the world economy has a two-tier class structure internal to global cities: People who live there are increasingly either highly paid professionals—lawyers, accountants, executives—or low-wage workers such as janitors, waitresses, and truck drivers.

Viewing the ramifications of advances in information technology more broadly, it can be argued that geography, borders, and time zones are all becoming irrelevant in the way people conduct their daily lives. Frances Cairncross calls this the "death of distance" and predicts that it will be the single most important economic force shaping society over the twenty-first century.[20] Among the most significant trends that she identifies are:

- Location will no longer be important to most business decisions. Companies will be able to locate anywhere on Earth.
- Because distance will no longer determine the cost of communicating electronically, companies will organize work in three shifts based on the world's three main time zones.

- The line between work and home life will blur as more people work from home.
- Small companies will be able to offer services that previously only very large companies could provide.
- New ideas and information will travel ever faster to the furthest corners of the world.
- Citizens will have greater freedom to locate anywhere and earn a living.

In light of these prognostications, William J. Mitchell foresees the creation of cities that encompass virtual places in cyberspace as well as physical ones—what he calls an "e-topia."[21] The global information network, Mitchell contends, is not just a delivery system for e-mail, the Internet, and digital television. Rather, it is a whole new form of infrastructure that will change the forms of our cities just as railroads, highways, and telephones did in the past. By rewiring hardware, replacing software, and reorganizing network connections, the new settlement patterns of the twenty-first century will be able to accommodate far-reaching configurations of electronic meeting places as well as decentralized production and distribution systems. Furthermore, as Mitchell sees it, the power of physical place will still prevail. As he explains: "Physical settings and virtual venues will function interdependently, and will mostly complement each other. . . . Sometimes we will use networks to avoid going places. But sometimes, still, we will go places to network."[22]

THE ROLE OF SOCIETAL FACTORS

It can be seen, then, that technology drives the urbanization process by creating opportunities—but the extent to which these opportunities are exploited is a function of economic, cultural, demographic, and other societal factors. Of particular significance is America's transition in the 1800s from an essentially rural society to an industrial society, which fomented attitudes of antiurbanism. With the spread of industrialization and the growth of cities in the 1830s and 1840s, many of the nation's most prominent authors, poets, and philosophers expressed anxiety at the likely consequences. For example, Ralph Waldo Emerson wrote: "That uncorrupted behavior which we admire in animals and young children belongs to the . . . man who lives in the presence of nature. Cities force growth and make men talkative and entertaining, but they make them artificial."[23] Henry David Thoreau philosophized that "cities corrupt—nature restores." Others who expressed similar views include Edgar Allan

Poe, Nathaniel Hawthorne, Henry Wadsworth Longfellow, and Jane Addams, the founder of the field of social work.

In their book *The Intellectual versus the City*, Morton and Lucia White contend that the antiurban bias of American intellectuals has shaped attitudes of Americans for many years. A recent Gallup poll reported that only 19 percent of Americans would prefer to live in a city. Part of the difficulty in shaping public policy on urban problems is linked to the long-standing tradition of antiurbanism. Furthermore, it can be argued that the large-scale migration of middle-class households out of the central cities to outlying suburban areas is another manifestation of negative attitudes toward cities.

Still another aspect of U.S. culture that continues to affect the way cities develop is the tradition of privatism. Sam Bass Warner Jr. believes this is the most critical element for understanding the development of American cities.[24] According to Warner, privatism

> has meant that the cities of the United States depended for their wages, employment, and general prosperity on the aggregate successes and failures of thousands of individual enterprises, not upon community action. It has also meant that the physical forms of American cities, their lots, houses, factories ,and streets have been the outcome of the real estate market of profit seeking builders, land speculators, and large investors. And it has meant that the local politics of American cities have depended on their actors, and for a good deal of their subject matter, on the changing focus of men's private economic activities.[25]

Values of privatism have prevailed throughout U.S. history. More recently, Manuel Castells observed that values of privatism have played a special role in economic restructuring, which in turn has significantly impacted urban conditions.[26] Particularly significant have been the military-industrial defense buildup in the 1980s and the dismantling of the urban welfare state in the 1990s. These, according to Castells, have exacerbated uneven regional development across the country and have damaged already depressed urban regions.

Of particular interest in viewing American society is the propensity of its citizens for engaging in group activities of all kinds. When Alexis de Tocqueville, a Frenchman, visited the United States in the 1830s, it was this tendency as a cultural phenomenon that most impressed him. "Americans of all ages, all stations in life and all types of disposition," he observed,

> are forever forming associations. There are not only commercial and industrial associations in which all take part, but others of a thousand different types—religious, moral, serious, futile, very general and very limited, im-

mensely large and very minute. . . . Nothing, in my view, deserves more attention than the intellectual and moral associations in America.[27]

As previously noted, urban observers such as Jane Jacobs have contended that the quality of public life is indeed influenced by norms of civic engagement. Jacobs noted that "social capital"—a term of which she is one of the inventors—is what most differentiated safe cities from unsafe and disorganized ones.[28] She argues that where cities are designed to promote networks of civic engagement, they benefit from lower crime rates and more stable neighborhoods, for such networks enable people to commit themselves to each other and to knit the social fabric.

In his book *Bowling Alone,* Robert Putnam noted that on a range of indicators of civic engagement including voting, political participation, newspaper readership, and participation in local associations, social capital in America was in decline.[29] In examining the possible reasons for this decline, he found that time pressures on two-career families were only marginally significant. So, too, was residential mobility, which has been in decline for the last half-century. More significant is suburban sprawl, because of which people must travel further to work, shop, and partake in leisure activities; consequently, there is less time available to become involved in groups. Electronic entertainment, especially television, is another factor; the time people spend watching television serves to drain involvement in group activities that contribute to social capital. Putnam estimates that up to 40 percent of the decline in group participation is linked to this phenomenon. However, Putnam contends that generational change is the most significant factor. As he explains, a "long civic generation," born in the early twentieth century, is now passing from the American scene. Their children and grandchildren are much less engaged in most forms of community life. As a rule of thumb, Putnam concludes that this factor might account for perhaps half of the overall decline.[30]

American society is changing in still another important way. This is based on what Richard Florida calls the "rise of the creative class."[31] The transformation of the U.S. economy from an industrial focus to one that is increasingly based on advanced technology now requires a class of professionals that can harness its full economic potential. According to Florida, the core of this class includes scientists, engineers, university professors, poets and novelists, artists, entertainers, actors, designers, architects, and think-tank researchers. Drawing on complex bodies of knowledge, these people engage in creative problem solving. He estimates that some 38 million Americans—30 percent of all employed people—belong to this class. As such, they add creative value.

Florida argues that the key to economic growth in communities depends on their ability to attract such persons. It is not so much a question of how *companies* decide where to locate, but rather how *people* decide to do so. Florida's theory is that "regional economic growth is driven by the location choices of creative people . . . who prefer places that are diverse, tolerant and open to new ideas."[32] Thus cities like Buffalo, Providence, and Memphis continue to lose members of the creative class to places like Austin, Boston, Washington, D.C., and Seattle.

THE STRUCTURE OF THE BOOK

As we have noted, the main focus of this book is the relationship between technological innovation and urban change. It explores how technology has transformed cities in the past and how it is likely to affect cities in the future. Urban-watchers and futurists are predicting radical change in urban life and form as the newest technologies place their imprints on cities. Among the questions posed are:

- What happens to cities in the shift away from an economy based on the production and distribution of material goods to one based more and more on the circulation and consumption of informational goods?
- How are cities and their surrounding regions being affected physically by advances in technology?
- What do these changes portend for the way cities are to be managed and governed?

Optimists view anticipated changes as the solution to many of the social and environmental ills of industrial society. Pessimists, on the other hand, paint a grim picture of a depressed urban era dominated by sinister forces that use technology for ominous purposes. Drawing from a broad array of scholarly analysis and observation, the present work attempts to clarify such speculation through careful examination of where we have been and, more important, where we are going as an urban society.

Continuing the discussion begun in this introductory chapter, chapter 2 reviews the development of technology in America as based on the emergence of new power sources: initially animate; then wind and water power; then driven by steam, electricity, and internal combustion; and finally, enabled by superconductivity (see table 1.1). Each of these, in turn, has made possible new technical applications in the areas of manufacturing, transportation, infrastructure, and communications. Over time, as we shall see, technological breakthroughs in these areas have dramatically altered the urban environment. This is discussed in the context of transitions from the preindustrial city to the industrial city and currently to the

Table 1.1.   Elements of Technology and Urban Change

Sources of Power	Applications	Urban Transformation
Animate		
Wind and water		
Steam		
Electricity		
Internal combustion		
Superconductivity		
	Production/manufacturing	
	Transportation	
	Infrastructure	
	Communications/	
	telecommunications	
		The preindustrial city
		The industrial city
		The postindustrial city

evolving postindustrial city. Chapter 3 presents a more in-depth view of communications technology and the changing role of cities as hubs or outposts for telecommunications and telematic networks. Chapter 4 discusses the implications of such transformation, focusing on the emergence of new urban forms from global cities to innovation centers. Finally, chapter 5 elaborates on the unfolding of the new urban paradigm and the strategies by which cities adapt to postindustrial forces of change.

NOTES

1. H. G. Wells, "The Discovery of the Future," *Nature* 65, no. 1684 (6 February 1902): 328.

2. Lewis Mumford, *The Culture of Cities* (New York: Harcourt, Brace, 1938).

3. Mumford, *Culture of Cities*, 393.

4. Mumford, *The Highway and the City* (New York: Harcourt, Brace & World, 1963), 222.

5. Le Corbusier, *The City of To-morrow and Its Planning* (Cambridge, Mass.: MIT Press, 1971).

6. Jane Jacobs, *The Death and Life of Great American Cities* (New York: Vintage, 1961).

7. Jacobs, *Death and Life*, 4.

8. See Mellier Scott, *American City Planning* (Berkeley: University of California Press, 1969), 33.

9. Marshall McLuhan, *Understanding Media: The Extensions of Man* (New York: McGraw-Hill, 1965).

10. Marshall McLuhan and Bruce R. Powers, *The Global Village: Transformations in World Life and Media in the 21st Century* (New York: Oxford University Press, 1989).

11. Joshua Meyrowitz, *No Sense of Place: The Impact of Electronic Media on Social Behavior* (New York: Oxford University Press, 1985).

12. Everett M. Rogers, *Diffusion of Innovations*, 3rd ed. (New York: Free Press, 1983).

13. Alvin Toffler, *The Third Wave* (New York: Morrow, 1980).

14. Tarik A. Fathy, *Telecity: Information Technology and Its Impact on City Form* (New York: Praeger, 1991).

15. William H. Dutton, Jay G. Blumler, and Kenneth L. Kraemer, eds., *Wired Cities: Shaping the Future of Communications* (Boston: G. K. Hall, 1988).

16. Toffler, *Third Wave*, and Alvin Toffler and Heidi Toffler, *Creating a New Civilization: The Politics of the Third Wave* (Atlanta: Turner, 1994).

17. Daniel Bell, *The Coming of Post-Industrial Society: A Venture in Social Forecasting* (New York: Basic Books, 1976); Alain Touraine, *The Post-Industrial Society; Tomorrow's Social History: Classes, Conflicts and Culture in the Programmed Society*, trans. Leonard F. X. Mayhew (New York: Random House, 1971).

18. John Naisbitt, *Megatrends* (New York: Warner Books, 1982).

19. Saskia Sassen, *The Global City: New York, London, Tokyo* (Princeton, N.J.: Princeton University Press, 1991).

20. Frances Cairncross, *The Death of Distance: How the Communications Revolution Will Change Our Lives* (Boston: Harvard Business School Press, 1997).

21. Willam J. Mitchell, *E-topia* (Cambridge, Mass.: MIT Press, 1999).

22. Mitchell, *E-topia*, 155.

23. Morton White and Lucia White, *The Intellectual versus the City, from Thomas Jefferson to Frank Lloyd Wright* (Cambridge, Mass.: Harvard University Press, 1964), 40.

24. Sam Bass Warner Jr., *The Private City: Philadelphia in Three Periods of Its Growth* (Philadelphia: University of Pennsylvania Press, 1968).

25. Warner, *Private City*, 4.

26. Manuel Castells, *The Informational City: Information Technology, Economic Restructuring, and the Urban-Regional Process* (Oxford, England: Blackwell, 1989).

27. Alexis de Tocqueville, *Democracy in America*, ed. J. P. Mayer, trans. George Lawrence (Garden City, N.Y.: Doubleday, 1969), 513–17.

28. Jacobs, *Death and Life*, 56.

29. Robert D. Putnam, *Bowling Alone: The Collapse and Revival of American Community* (New York: Simon & Schuster, 2000).

30. Putnam, *Bowling Alone*, 283.

31. Richard Florida, *The Rise of the Creative Class: And How It's Transforming Work, Leisure, Community and Everyday Life* (New York: Basic Books, 2004).

32. Florida, *Rise of the Creative Class*, 223.

2

The Transformation of
Urban America

Borrowing from the Tofflers,[1] who write of the transformation of American society through three "waves" of technological change, this chapter takes a closer look at how U.S. cities have developed from preindustrial times, through industrialization, to what is now being viewed as the postindustrial era—what the Tofflers call the Third Wave.

The preindustrial era, or First Wave, was rooted in the agricultural revolution and characterized by the transition from hunting and gathering to food cultivation and the domestication of animals. Today, First Wave technology has just about spent its force and is no longer dominant in the United States or other Western countries. The Second Wave of change was generated by the industrial revolution, and its momentum is still felt in most countries where factories, steel mills, ports, and railroads are the mainstays of the national economies. But even as the process of industrialization continues, a postindustrial Third Wave is now transforming Western societies at an accelerating rate, making everything that it touches obsolete. Whereas Second Wave sectors rely on manufacturing and mass production, Third Wave sectors ascend to dominance via new ways of creating and exploiting knowledge: they invest in innovation and trade in information, software, and advanced technology. The Third Wave began to gather strength in the United States in the 1950s. Since then, it has made itself felt in other industrial countries, including Germany, Japan, England, and France. Moreover, some other so-called developing nations such as India, China, and Singapore are also making rapid progress as high-tech societies.

Table 2.1.   Leading Cities in the United States, 1790

Rank	City	Rank	City
1	New York, N.Y.	11	Portsmouth, N.H.
2	Philadelphia, Pa.	12	Brooklyn, N.Y.
3	Boston, Mass.	13	New Haven, Conn.
4	Charleston, S.C.	14	Taunton, Mass.
5	Baltimore, Md.	15	Richmond, Va.
6	Salem, Mass.	16	Albany, N.Y.
7	Newport, R.I.	17	New Bedford, Mass.
8	Providence, R.I.	18	Beverly, Mass.
9	Gloucester, Mass.	19	Norfolk, Va.
10	Newburyport, Mass.	20	Petersburg, Va.

Source: Reprinted by permission of the National Council for Geographic Education, from James E. Vance Jr., "Cities in the Shaping of the American Nation," *Journal of Geography* 75 (1976): 41–52.

THE PREINDUSTRIAL CITY, 1760–1820

In contrast to Europe and Asia, most American cities are relatively young and were formed in the course of industrialization, beginning around the 1820s. Prior to that, in the preindustrial period, U.S. cities were "specks in the wilderness."[2] There was no city of national scope, and urban clusters such as Philadelphia, New York, Boston, and Charleston were relatively small, with populations of less than thirty thousand. Table 2.1 shows that at the time of the 1790 census, almost all major population centers were ports along the Atlantic coast. These ports were backed up by small hinterlands and were oriented toward the sea and Europe; consequently, they could be considered part of the West European urban network.

From roughly the 1790s to the 1820s, there was little evidence of change in the urban landscape. Some growth in inland centers occurred along waterways that began to penetrate the interior—primarily the Erie Canal, the lower Great Lakes, and the Ohio River. Most inland roads were, at best, patchy. They usually consisted only of animal and cart tracks, which became rutted and worn. In wet weather, many were impassable. Thus, any form of vehicular traffic was difficult to maintain, and a great deal of traffic went by horseback, carrying goods in large baskets. It is understandable, therefore, that commerce and trade emerged in seaboard and river cities, where boats provided effective transport. Improvements in ship construction and the growing mastery of sailing techniques produced a remarkably versatile form of maritime transport that required no fuel after the initial fitting out. Such transport could carry a substantial cargo over the seas to Europe or to the Caribbean islands.

The absence of any breakthroughs in land transportation during this time meant that Eastern Seaboard cities continued to function primarily as maritime centers with no appreciable increase in size. The typical preindustrial city was a settlement no more than a mile in radius, hence small enough to be bound together by personal networks of family, friendship, or business. Social organization was based largely on primary relations and consisted of frequent personal communication and face-to-face contact. Only a primitive specialization with respect to urban land existed, and neighborhoods tended to reflect a mix of economic and social groupings with little clustering along ethnic or occupational lines.

Manufacturing was tied primarily to hand tools, water power, and animal-driven machines. Given the state of technology, American factories were initially distributed around the countryside and became the basis for small communities. An observer of American conditions during this period describes how small towns evolved wherever there were good mill sites.

> To this mill, the surrounding lumberers, or fellers of timber bring their logs, and either sell them, or procure them to be sawed into boards or into plank, paying the work in logs. The owner of the sawmill becomes a rich man, builds a large wooden house, opens a shop, denominated a store, erects a still, and exchanges rum, molasses, flour and port, for logs. As the country has by this time begun to be cleared, a flour-mill is erected near the saw-mill. Sheep being brought upon the farms, a carding machine and fulling-mill follow . . . the mills becoming every day more and more a point of attraction, a blacksmith, a shoemaker, a tailor, and various other artisans and artificers, successively assemble. . . . So a settlement, not only artisans, but farmers is progressively formed in the vicinity; this settlement constitutes itself a society or parish; and, a church being erected, the village, large or smaller, is complete.[3]

By the 1760s, colonial merchants had begun to invest in modest manufacturing enterprises. In New England, swift-running streams or falls contributed to the growing use of water power in production and spurred the spread of the textile industry. In Boston, craftsmen were refining their skills in the manufacture and export of furniture and shoes; and shipbuilding continued to expand as an investment outlet in that city as well as in New York and Philadelphia. In several towns, merchants formed organizations for the purpose of developing manufacturing as a way of diversifying the local economy and enhancing trade. For example, in 1776, Philadelphia merchants organized the United Company of Philadelphia for Promoting American Manufactures; and somewhat earlier, in 1765, New York merchants formed the Society for the Promotion of Arts, Agriculture, and Economy.

For the most part, however, cities functioned as market centers engaged in the import and export of various commodities. Oriented toward the waterfront, the urban market could usually be found at the water's edge, where an array of wares and foods provided an impressive display for the shopper. Finding marketable commodities to ship abroad was a pressing need for urban merchants. When Philadelphia's fertile countryside began to produce wheat in the eighteenth century, it became a major source of trade with Caribbean and European ports; and by 1775 the city exported more flour to these ports than did all the other colonies combined. Boston initially relied on the fur business as a main source of trade. Because animals were a finite commodity and security on the frontier was precarious, however, Boston merchants soon turned to exporting fish. When the fish trade proved marginal, they turned to the illegal loading of molasses in the West Indies, which was forbidden under English mercantile laws.

Taking a closer look at Philadelphia in the late eighteenth century, Sam Bass Warner Jr. characterized the city as a "private city" with a privatistic ethic.[4] The great majority of the 23,700 residents were independent shopkeepers and artisans. Merchants engaged in production and trade were generally accorded higher social status, while a small number of indentured servants, slaves, and hired servants labored at the bottom of the social system. As the city grew and commerce became the dominant feature of economic life, the gap between rich and poor tended to widen. Wealth, however, was not confined to a closed aristocracy, because there was movement within and among the various class categories. The availability of credit and of opportunity for breaking into international trade enabled those with entrepreneurial ambitions to move into and out of the mercantile class with great frequency and facility.

Political values in Philadelphia were also privatistic, as represented by the fact that citizens did not want an activist government. The only legitimate functions of city government were to manage the market and administer the local records court. There was no public water supply, and most streets were unpaved. Independent commissions of assessors, street commissioners (city wardens in charge of the night watch and street lighting), and a board of overseers for the poor actually accomplished little. Instead, Warner notes that most things got done through informal networks of community leaders who functioned according to informal rules.

THE EARLY INDUSTRIAL CITY, 1820–1870

The most notable change of the nineteenth century from the 1820s to the 1870s was the sudden increase in the population of cities. In 1790, approximately 5 percent of the nation's population had lived in cities. Thirty years later, the urban population had increased only slightly, to 7 percent, and only twelve cities had populations large than ten thousand. But thereafter, in every decade from 1820 to 1870, the urban population grew three times as fast as the national population. By 1860, more than one hundred cities had populations larger than ten thousand, and eight of those had more than one hundred thousand residents. One city—New York City—grew beyond one million. By 1870, furthermore, roughly one out of four Americans lived in cities.[5]

At this same time, hand tools and horse-powered machines were further perfected, and water power and steam engines came into greater use. The year 1825 saw a major technological breakthrough with the completion of the Erie Canal, which linked the Great Lakes and the Atlantic Ocean at New York City. Thereafter, the same ton of flour that had taken three weeks to move from Buffalo to New York at a cost of $120, made the trip in eight days at a cost of $6, and New York almost overnight turned into an economic powerhouse. In Canada, the opening of the Welland Canal in 1829 linked Lake Erie to Lake Ontario, thereby providing a water route to the west that helped to counter the economic threat of the United States' Erie Canal.

The introduction of steam power created major transportation routes on the western rivers and resulted in the expansion of ports on the inland waterways. By the 1830s, the buildup of steamboat tonnage on the Ohio-Mississippi-Missouri system served to increase the tonnage of general-cargo vessels on the Great Lakes.

Rail mileage also grew rapidly. In the six years after 1830, when the first rail line went into operation, more than one thousand miles of track were laid. Construction of the first railway in the United States was begun on July 4, 1828, in Baltimore. Baltimore had grown slightly larger than Philadelphia, and with 80,600 inhabitants it was second only to New York among American cities. The rail line would link Baltimore to the Ohio Valley, opening it to western trade just as New York had tapped the Great Lakes region with the Erie Canal. Although the Baltimore & Ohio Railroad did not reach the western terminus of the line at Wheeling, Virginia, until 1853, the completion of sections before that date gave evidence that this form of transportation technology could be profitable.

Prior to 1850, the railroad, along with the canal system, promoted intraregional trade at the same time that it furthered the economic prospects of eastern coastal cities such as Baltimore and Philadelphia. By 1835, Boston merchants had completed rail lines to Lowell, Providence, and Worcester. These cities, in turn, built branches to smaller localities, tying the entire system to Boston and, consequently, to other Atlantic ports, which fed into international commerce.

The growth of western cities was very similar to the growth of the urban Northeast. Before 1820, commerce and trade among the fledgling western cities were severely constrained by poor transportation. This changed in the 1830s when canals were constructed connecting cities to the Great Lakes. Cleveland, Toledo, and Chicago prospered as a result and grew rapidly with the introduction of the steamboat, which could navigate through the swift upstream waters of the Ohio and Mississippi rivers. The railroad and the steam-powered locomotive, however, were the ultimate weapons for western development. More important, they made possible the development of a national transportation system through the integration of major waterways and regional rail webs. By the 1850s, the railroad had successfully reoriented the continental trade pattern from a north–south to a west–east direction. These changes spurred the growth of coastal ports with large harbors at the same time that they hurt the economy of neighboring small ports.

Undoubtedly the high point in the development of railroad technology was the building of the first transcontinental railroad.[6] The idea of a rail line from the Atlantic to the Pacific had lingered for a long time—it was considered by Congress in the late 1840s and refined into a more detailed plan in the 1850s. Most observers doubted that the project would be realized very soon, given the technological difficulties and the huge expense. Yet despite the constraints, the transcontinental railroad was constructed from New York to San Francisco with great dispatch. In 1865, abetted by land grants from Congress, railroads in California and Nebraska began building toward each other. As incentive, each was trying to maximize its share of the federal subsidies given for every mile of track laid. The culmination of four years of frenzied work came at Promontory Summit, Utah, on May 10, 1869, when the line was completed with the driving of an allegedly golden spike.

In addition to its utilization in water and land transportation, steam power was also used in manufacturing during this period, but its impact was generally limited because iron rail lines of the time and light transport equipment could not accommodate long hauls of coal or other heavy commodities. Consequently, in the 1870s local waterpower sites still influenced the location of industry, and waterwheels were still providing about half of the inanimate energy for manufacturing. As can be seen in table 2.2, the greater number of leading U.S. cities at this time were still the coastal and river cities located primarily in the eastern part of the country. However, the rapid growth of such cities as Chicago, Milwaukee, Minneapolis, Detroit, Cleveland, and St. Louis signified the emergence of an inland manufacturing belt. The rise of San Francisco on the West Coast was also notable.

Table 2.2.   Leading Cities in the United States, 1870

Rank	City	Rank	City
1	New York, N.Y.	11	Pittsburgh, Pa.
2	Philadelphia, Pa.	12	Buffalo, N.Y.
3	Brooklyn, N.Y.	13	Washington, D.C.
4	St. Louis, Mo.	14	Newark, N.J.
5	Chicago, Ill.	15	Louisville, Ky.
6	Baltimore, Md.	16	Cleveland, Ohio
7	Boston, Mass.	17	Jersey City, N.J.
8	Cincinnati, Ohio	18	Detroit, Mich.
9	New Orleans, La.	19	Milwaukee, Wis.
10	San Francisco, Calif.	20	Albany, N.Y.

Source: Reprinted by permission of the National Council for Geographic Education, from James E. Vance Jr., "Cities in the Shaping of the American Nation," *Journal of Geography* 75 (1976): 41–52.

THE INDUSTRIAL CITY COMES OF AGE, 1870–1920

From roughly the 1870s to the early 1900s, urbanization as a product of industrialization had become a dominating factor in the life of the nation. Undoubtedly, the event that best symbolized the coming of age of the industrial city was the Great Philadelphia Centennial Exposition of 1876, which celebrated the hundredth anniversary of the city's participation in the Declaration of Independence. The centerpiece of the exhibition was Machinery Hall, where visitors marveled at the demonstrations of sewing machines, assembly-line shoe production, automatic cornhuskers, and power drills for digging holes. They inspected state-of-the-art harvesters and plows and watched wood pulp being made into paper. There was even a machine to make hens lay eggs. At the center of it all, rising thirty feet into the air, loomed a six-hundred-ton Corliss steam engine that ran all of the exhibition's four thousand machines.[7]

Machines such as the Corliss steam engine symbolized the new factory system that was moving society toward new forms of industrial concentrations. These concentrations were a product of the growing demand by new corporations for mass-production facilities at the relatively few sites that could deliver economies of scale by utilizing greater specialization of labor. Compared to those built before 1880, these new factories were huge. Many had their own rail lines, streets, and power stations. Where the old factories had blended into the townscape, the new steam-driven facilities were visibly distinct.

As demonstrated by the Corliss engine, steam power was literally the driving force in the development of the industrial city. The steam engine had become the main provider of industrial power—as the source of pumping power for mines and waterworks, as the means of moving locomotives, as the energizing force of machinery, as the propulsive force of steamships, and, with the invention of the steam turbine, as the generator of electricity.

Other important technological breakthroughs occurred during this period. The first commercial production of Bessemer-process steel in America began moving steel products into the world market. By the 1870s, steel rail lines were replacing iron ones. This allowed for more powerful locomotives that could travel at higher speeds and haul bulk goods over longer distances. The greater availability of coal carried by long-haul rail was soon complemented by the availability of central-station electric power in the 1880s. Rail-gauge and freight-car parts were standardized to make possible interline exchange and coast-to-coast shipments. These developments sounded the death knell for the small waterpower sites, which were now giving way to centralized metropolitan rail centers and giant markets with their superior accessibility.

Between the end of the Civil War and the dawn of the twentieth century, American inventors filed nearly a half-million patents, transforming the U.S. economy and the role that cities would play in that economy. Alexander Graham Bell demonstrated his telephone, making possible instantaneous communication over long distance. Thomas Edison opened his laboratory in Menlo Park, New Jersey, where he demonstrated the phonograph and the electric light bulb. Refrigerated rail cars were introduced, ushering in a new era of regional specialization in agriculture and centralization of the packing industry at major rail centers. Innovations in mass transit such as electrified trolley cars served to push development to the outer sectors of urban settlement. Ongoing refinements in infrastructure facilities—ranging from street surfacing and public lighting to water and sewerage systems (see box 2.1)—made high urban densities safer.

By the 1880s, the wonders of electricity were being demonstrated in expositions that previously would have closed at nightfall. At the Louisville Southern States Exposition of 1883, the Edison Electric Lighting Company installed 4,600 lightbulbs, creating a sensation. This was the last exposition that used gas for illumination. A few years later in 1894, the Columbian Exposition in Chicago showed more lighting than any city in the country. A spectacular view was provided from atop another technological marvel invented for the Chicago Fair: the Ferris Wheel, itself studded with lightbulbs.[8]

The industrial city that emerged from this development was huge in size and densely settled at the core, and it contained a highly diverse population derived from the inflow of migrants from rural areas and massive waves of mostly European immigrants. Whereas in 1860 only sixteen

Box 2.1. Building the Municipal Infrastructure

City officials had long appreciated the importance of adequate water supplies. Fear of epidemics and fires, coupled with the pollution of wells by seepage from graves and privies, forced city fathers to tap new sources and often to bring in water from outside municipal boundaries. Philadelphia built a public waterworks system in the 1790s. New York, Boston, and a few other large cities followed suit some forty years later. Various physicians and public health reformers advertised the sanitary reasons to provide not only water but pure water. But who should construct and administer the new systems, the private sector or government?

Until the 1850s, most cities relied upon private firms, but in most cases the private efforts failed technically and economically, especially in the largest cities. Few private entrepreneurs or corporations had the capital, the condemnation power, the concern for public health, or the economic will to build and maintain water supply systems that would serve the entire public. Concerned about profits, few private companies proved to be willing to serve poorer people from whom they could expect meager revenues. Thus, municipal ownership and administration gained slowly between the 1860s and the 1890s. By the turn of the century, however, only nine of the fifty largest cities still had privately owned water supplies. By 1910, more than 70 percent of cities over 30,000 population owned their waterworks. Over the last half of the century, city officials discerned sanitary, technological, and political reasons to provide this service at municipal expense and under municipal control.

The addition of adequate water did not end environmental pollution. To some extent, it increased it, for now cities had to dispose of vast quantities of water brought in by the aqueduct. Existent surface drainage was inadequate. The new water closets of the 1860s and 1870s overflowed the old privy waste disposal systems, soaked the urban water tables, and converted large portions of city land and streets into stinking morass. Once again the solution was physical and technical. During the 1870s and 1880s, city leaders undertook expensive programs of sewer building. They also began massive paving programs to improve drainage and to cover the wastewater-saturated soil of urban streets. The engineers who shepherded these projects emphasized their sanitary functions nearly as much as their traffic-bearing functions.

Source: Reprinted by permission of the Organization of American Historians from Stanley K. Schultz and Clay McShane, "To Engineer the Metropolis: Sewers, Sanitation, and City Planning in Late Nineteenth-Century America," *Journal of American History* 65 (September 1978): 39.

cities had more than 60,000 people, eleven cities could boast of having more than 250,000 by the end of the century. The demand for war material during the Civil War, combined with advancement in manufacturing, brought about dramatic changes in the industrial base of the city.

In 1860, only two cities, New York and Philadelphia, had more than 25,000 manufacturing employees (see table 2.3). By the dawn of the

Table 2.3.   Manufacturing Employment in Major Cities, 1860–1900 (thousands of employees)

1860		1880		1900	
New York	106	New York	282	New York	511
Chicago	—	Chicago	79	Chicago	298
Philadelphia	99	Philadelphia	186	Philadelphia	266
		St. Louis	42	St. Louis	93
		Boston	59	Boston	81
		Baltimore	56	Baltimore	85
		Pittsburgh	41	Pittsburgh	98
		Cleveland	—	Cleveland	64
		San Francisco	28	San Francisco	46
		Detroit	51		
		Louisville	33		

Note: Cities with more than 25,000 manufacturing employees are listed, ordered according to 1900 population size.
Source: Reprinted by permission of the publisher from Allan R. Pred, *The Spatial Dynamics of U.S. Urban Industrial Growth, 1800–1914* (Cambridge, Mass.: MIT Press, 1966), 114.

twentieth century, more than 90 percent of all production took place in cities. The bulk of manufacturing initially developed in the Northeast and the Midwest. Subsequently, manufacturing began to grow faster in the interior of the country than in older coastal cities. In many instances, cities that served as rail hubs became important industrial centers.

The City of Chicago provides useful illustration. Using the criterion of valued added (defined as the difference in the dollar cost of producing a product and the value of the finished product), industrial production in that city soared from $5 million in 1860 to $283 million in 1890, placing it third after New York and Philadelphia (see table 2.4). As shown in table 2.3, the number of industrial workers in Chicago increased rapidly in a twenty-year span—from 79,000 to nearly 300,000. In addition, table 2.4 shows that four other cities—New York, Philadelphia, St. Louis, and Boston—grew rapidly, exceeding $100 million in value added over a thirty-year span. Several cities with smaller initial bases also evidenced significant growth—notably, Pittsburgh, San Francisco, Cleveland, and Detroit.

Compared to the preindustrial city, which was basically a marketplace and the seat of commerce and trade, the industrial city—particularly its business and industrial districts—came to be viewed as a great economic machine. Here land, labor, raw materials, and business organization were brought together as critical components in the production process. From a broad historical view, Arthur M. Schlesinger describes the pluses and minuses of this phenomenon:

Table 2.4. Value Added by Manufacturing in Major Cities: 1860, 1890, and 1982

	1860		1890		1982	
	$ millions	Rank	$ millions	Rank	$ millions	Rank
New York	90	1	612	1	61,548	1
Philadelphia	70	2	298	2	21,161	2
Chicago	5	7	283	3	36,363	3
St. Louis	9	4	120	4	—	—
Boston	18	3	118	5	17,529	7
Baltimore	9	5	76	6	—	—
Pittsburgh	6	6	64	7	—	—
San Francisco	2	8	64	8	26,315	4
Cleveland	2	9	53	9	12,804	10
Detroit	2	10	40	10	18,381	6
Los Angeles	—	—	—	—	54,168	2
Houston	—	—	—	—	16,946	8
Dallas	—	—	—	—	14,691	9

Source: Reprinted by permission of the publisher from Allan R. Pred, *The Spatial Dynamics of U.S. Urban Industrial Growth: 1800–1914* (Cambridge, Mass.: MIT Press, 1966), 114.

In its confines were focused all the new economic forces: the vast accumulations of capital, the business and financial institutions, the spreading railway yards, the gaunt smoky mills, the white-collar middle classes, the motley wage-earning population. By the same token the city inevitably became the generating center for social and intellectual progress. . . . In the city were to be found the best schools, the best churches, the best newspapers, and virtually all the bookstores, libraries, art galleries, museums, theaters and opera houses. It is not surprising that the great cultural advances of the time came out of the city, or that its influence should ramify to the farther countryside.[9]

THE MASS-PRODUCTION METROPOLIS, 1920–1970

The industrial city of the late nineteenth and early twentieth centuries was a city of concentration and centralization. Since steam is generated more cheaply in large quantities and must be used close to where it is produced, steam power fostered a compact city. Manufacturing was usually located in a core area that surrounded the central business district within easy reach of rail and water transportation. This, in turn, tended to concentrate workers, managers, and wholesale distributing activities near the factories. Because the transportation technology of the time was limited, workers had to live near the factories. This gave rise to densely packed tenement housing, because only the very wealthy could afford the daily commute from the suburbs.

During the 1920s and 1930s, businesses found it necessary to locate their offices in close proximity to each other to facilitate the flow of information. High real estate values in the central business district were an inevitable consequence of business's demand for a central location. The downtown core became even more densely settled as inventions such as the elevator and the steel girder allowed buildings to rise to ever greater heights. The centers of the largest cities changed in two basic ways that distinguished them from the center of small communities: The new downtowns expanded out and up, driving residential property out of the center and creating a skyline of tall monuments to business and finance; and the center generated a mix of functions—retail, corporate, and entertainment—each appealing to different types of users.

In the nineteenth century, separation of place of residence from place of work was a luxury restricted to the wealthy few. Common people either walked or rode the horse-drawn streetcars to work. At the beginning of the twentieth century, the typical New Yorker lived approximately two blocks (roughly a quarter of a mile) from his or her place of work. Half the population of Chicago lived within 3.2 miles of the center of the city. Later, as the electric streetcar came into greater use, more and more people could commute relatively rapidly and inexpensively from the outer areas of the city to the center. This led to the establishment of residential suburbs in strips along the streetcar right-of-way. Land lying between the streetcar lines usually remained undeveloped, though commercial districts often emerged along the streetcar line at the intersection of major avenues. Consequently cities came to manifest a star-shaped configuration, with the linear rail lines forming the points of the star. The cities would maintain this shape until automobile transportation began to dominate.

By 1920, the growing use of automobiles, trucks, and telephones increased the movement of people, goods, and ideas. The automobile provided greatly expanded mobility to the average urban dweller and made possible the rapid settlement of areas on the periphery of the central city (see box 2.2).[10] Automobile registration in the United States increased from 2.5 million in 1915 to 9 million in 1920 and 26 million in 1930. By 1927 more than half the families in urban America owned an automobile, compared to fewer than one out of six in 1920. Henry Ford had opened his vast new plant outside Detroit in 1919, and five years later, using assembly-line technology, he was able to drop the price of the popular Model T from $959 in 1919 to $290 in 1924.

Federal government subsidies for highway construction served to accelerate the trend toward automobile transportation. The Federal Road Act of 1916 gave grants to states that organized highway departments, and the Federal Road Act of 1921 designated two hundred thousand roads as "primary" and thus eligible for matching federal funds. The 1921

Box 2.2. The Automobile Revolution

[Henry] Ford's energetic driving down of prices helped to make the automobile more popular, but equally responsible were a series of vital improvements: the invention of an effective self-starter, first designed by Charles F. Kettering and installed in the Cadillac in 1912; the coming within the next two or three years of a demountable rim and the cord tire; but only 2 percent of the cars manufactured in the United States were closed; by 1926, 72 percent of them were.

What had happened was that manufacturers had learned to build closed cars that were not hideously expensive, that did not rattle themselves to pieces, and that could be painted with a fast-drying but durable paint; and that meanwhile the car-buying public had discovered with delight that a closed car was something quite different from the old "horseless carriage." It was a power-driven room on wheels—storm-proof, lockable, parkable all day and all night in all weathers. In it you could succumb to speed fever without being battered by the wind. You could use it to fetch home the groceries, to drive to the golf club or the railroad station, to cool off on hot evenings, to reach a job many miles distant and otherwise inaccessible, to take the family out for a day's drive or a week-end excursion, to pay an impromptu visit to friends forty or fifty miles away, or, as innumerable young couples were not slow to learn, to engage in private intimacies. . . . And if the car was also a frequent source of family friction ("No, Junior, you are not taking it tonight"), as well as a destroyer of pedestrianism, a weakener of the churchgoing habit, a promoter of envy, a lethal weapon when driven by heedless, drunken, or irresponsible people and a formidable convenience for criminals seeking a safe getaway, it was nonetheless indispensable.

Source: Reprinted by permission of HarperCollins Publishers, Inc., from Frederick Lewis Allen, *The Big Change: America Transforms Itself, 1900–1950* (New York: Harper, 1952), 123.

---

law also created the Bureau of Public Roads to plan a highway system intended to connect cities of more than fifty thousand residents. At roughly the same time, state governments began placing a tax on gasoline as the means of paying for road construction and maintenance. When combined with private funds, government subsidies helped to create a national highway system by 1930.

In earlier times, business and industries needed access to railroad sidings and a labor pool, and both were available in the central cities. But by the 1920s, people were beginning to move to the suburbs in larger numbers. Downtown congestion, the growth of truck freight transport, and the increasing reliance on skilled workers lessened the attraction of central location. The introduction of high-voltage electrical systems, moreover, reduced the cost of power transmission to the outer areas of the metropolis. This meant that factories no longer had to be close to their power sources. Because manufacturing required large areas of ground space, four- and five-story factories located at rail and waterfront depots were soon judged obsolete, and manufacturers began to look to the suburbs where cheaper land was available. As factories moved out, taking workers with them, retailers began to follow.

As the airplane became the newest form of commercial transportation, airports were being located on the outskirts of cities. This construction accelerated during the 1930s when the airline industry began to expand. For growing numbers of airline passengers, the connection between transportation and downtown rail terminals began to weaken. Slowly but surely, the central city was losing many of the functions it had accrued at its peak in the late nineteenth century.

Some institutions, however, were inclined to remain in the city. While manufacturing was dispersing to suburban locations, the offices that administered and controlled the enterprises tended to remain in the central business district. In 1920 more than one-quarter of the nation's three hundred largest firms were located in New York City. Later, from the mid-1960s to the early 1970s, there was a more-than-50-percent increase in office space in older cities such as New York and Chicago, while Houston doubled its office space. Such functions as finance, government, management, and law have tended to remain at the center of the city, partly because they do not require great amounts of space per worker but, of greater significance, also because the compactness of the central business district facilitated informal face-to-face communications. Top management did not want to become isolated from the information networks that evolve when a number of firms in the same business or industry are located in the same general area.

To be noted as well, many tangential services were within easy reach of these same downtown locations. These included advertising, accounting, tax information, legal services, mailing, and even cultural events. Still another consideration was that corporations could achieve high visibility and advertisement for themselves by housing their headquarters in centrally placed skyscrapers that could be seen from great distances (see box 2.3). New York City and Chicago soon became the skyscraper capitals of the nation and the world, with such structures as the Woolworth Building (New York, built in 1913), the Chrysler Building (New York, 1929), the Chicago Tribune Tower (1923), the Chicago Board of Trade Building (1930), and the Empire State Building (New York, 1931). Later Chicago could boast of the Sears Tower (1974) as the tallest building in the world, and New York could vaunt the twin towers of the World Trade Building (1972) as the second tallest.

Brian Berry notes that prior to the 1950s the centrifugal movement of urban population was based largely on improvements in local trans-

Box 2.3. The Skyscraper as Symbol of the Central City

Corporations . . . found it necessary to demonstrate high visibility. More people, in other words, were likely to see a corporate headquarters downtown than at a low-cost suburban site. The corporate building became part of the advertising campaign for numerous large firms. Frank W. Woolworth demonstrated the efficacy of this strategy in 1913, and the skyscraper boom continued into the 1920s. The Chrysler Building, financed by auto tycoon Walter Chrysler and completed in 1929, was perhaps the most graceful of the gargantuan structures, a slender Art Deco cathedral of commerce rising seventy-seven stories above New York. The structure was capped by six stories of stainless steel featuring a frieze of abstract automobiles circling the tower. By this time Standard Oil magnate John D. Rockefeller had broken ground for what chic planners in the 1980s would call a multiuse development, a complex of office and retail buildings known as Rockefeller Center. The nation's major communications giant, the Radio Corporation of America, became the Center's primary tenant.

The skyscraper, which made its first appearance in the late nineteenth century, became the prime symbol for the twentieth-century city. Sleek, built of the latest materials, furnished with a fleet of fast elevators, ornamented with icons from the new technology, and creating a brand-new urban landscape and climate, the skyscraper fascinated tourists and residents alike, inspiring a generation of artists and photographers who sought to capture the excitement of the modern metropolis by focusing on its most evident symbol. Photographers such as Alfred Stieglitz, Edward Steichen, and Louis Hine took the skyscraper as their frequent subject, often resorting to impressionistic effects to soften and humanize the buildings.

Source: Reprinted by permission of the authors, from David R. Goldfield and Blaine A. Brownell, *Urban America: A History*, 2nd ed. (Boston: Houghton Mifflin, 1990), 299, 300.

portation and communication facilities as well as on the continued expansion of the business core.[11] Other uses were usually found for abandoned residential properties adjoining commercial and industrial districts; however, that kind of replacement began to decline in the later 1950s and had ceased entirely by the 1970s. Between 1950 and 1960, all of the large older cities lost population. Boston lost 13 percent while its suburbs grew by 17 percent. New York and Chicago lost less than 2 percent each, but their suburbs grew by over 70 percent. So, too, did the industrial base shrink. Between 1947 and 1967, the sixteen largest and oldest central cities lost an average of 34,000 manufacturing jobs each, while their suburbs gained an average of 87,000 manufacturing jobs. This trend continued through the 1970s.

Taking note of the growth of population and jobs in the suburbs, entrepreneurs during the 1950s and 1960s began transforming suburbia into self-sufficient communities independent of the central city. Merchants abandoned Main Street for strip locations along highways and for large shopping centers. For growing numbers of firms, the suburban office park had become the location of choice because it reduced commuter travel time for employees. Moreover, the most rapidly expanding segments of the economy—namely, electronics, chemicals, pharmaceuticals, and aeronautics—were being established in the suburban hinterland, leaving declining industries such as iron and steel, textiles, and automobiles to the old central city. The continuing outward movement of urban households and institutions was contributing to growing inventories of deteriorated and abandoned buildings.

THE POSTINDUSTRIAL METROPOLIS, 1970–PRESENT

By the late 1970s and early 1990s, it had become clear that the traditional concept of a single center dominating the metropolis was being displaced by a newer vision of multiple centers dispersed throughout the region. Today industry and commerce are spread throughout the urban landscape, and the metropolitan area functions as a total economy.

Underlying the rise of metropolitan-wide economies is the fact that present-day technology gives firms the freedom to locate anywhere within the metro area without loss of efficiency. Formerly, most central cities had advantages stemming from the need for firms to cluster at the core in order to reduce the costs of transportation and communication. However, many of these advantages have been eliminated as the importance of distance for these functions has been reduced. Information and communication technology have made locations of firms less significant. Instead, firms move to where costs are lower and quality of life higher; and jobs now move to where industries and people want to locate. Similarly, advancements in automobile technology and the spread of highways have meant that the people have come to measure distance in terms of time it takes to carry out their daily chores rather than in miles of travel (see box 2.4). As a result, older central cities continue to lose their hold on people and industry as they compete with suburban communities on the basis of such factors as available amenities, niche markets, and tax benefits.

From Urban Concentration to Suburban Deconcentration

Figure 2.1 shows that as of 2000, approximately 50 percent of the population was located in the suburbs. This compares to about 31 percent in 1960 and 23 percent in 1950. While rural (nonmetropolitan) areas have contin-

---

Box 2.4.   A City of Time Instead of Space

Even the largest of the old "big cities" had a firm identity in space. The big city had a center as its basic point of orientation—the Loop, Times Square—and also a boundary. Starting from the center, sooner or later one reached the edge of the city.

In the new city, however, there is no single center. Instead, as [Frank Lloyd] Wright suggested, each family home has become the central point for its members. Families create their own "cities" out of the destination they can reach (usually traveling by car) in a reasonable length of time. Indeed, distance in the new cities is generally measured in terms of time rather than blocks or miles. The supermarket is 10 minutes away. The nearest shopping mall is 30 minutes in another direction, and one's job 40 minutes away by yet another route. *The pattern formed by these destinations represents "the city" for that particular family or individual.* The more varied one's destinations, the richer and more diverse is one's personal "city." The new city is a city à la carte.

Source: Reprinted by permission of the author, from Robert Fishman, "America's New City: Megalopolis Unbound," *Wilson Quarterly* (Winter 1990): 38.

---

ued to lose population, the percentage of the population living in central cities continues to fluctuate around 30 percent. However, the overall figures for central city populations conceal important differences between relatively newer cities in the South and West that are experiencing growth (see table 2.5) and the many older industrial cities in the East and Midwest that are experiencing decline (see table 2.6).

In addition to losing population, the data in figure 2.2 show the extent to which the central cores of major urban areas have been losing manufacturing. Similarly, figures 2.3 and 2.4 show that retailing and wholesaling have also been departing from older central cities in the Northeast and Midwest. What these cities lose, their suburbs gain. For example, in the New York region, two-thirds of all retail is located in the suburbs. This trend can be explained by the fact that retailing follows the market: As more and more people—especially those with higher than average incomes—move to the suburbs, many consumer services have become predominantly suburban. At the same time, wholesaling has been moving to the suburbs to be near beltways and interstate highways and to take advantage of available lower land costs.

As people and firms continue to move to the suburbs, central cities and suburban communities reveal important demographic differences. First, large concentrations of poor people, many of whom are members of ethnic or racial minorities, remain in the cities. Table 2.7 shows that as of 2000, blacks and Hispanics made up 21 percent and 19 percent, respectively, of

Chapter 2

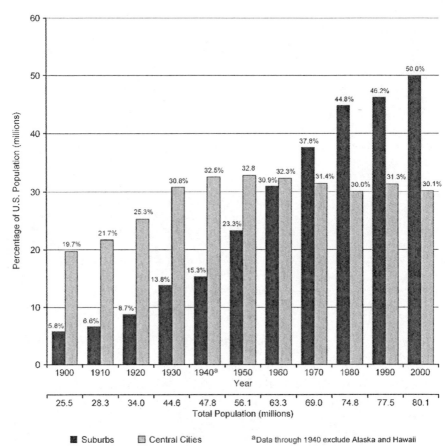

Figure 2.1. Percentage of U.S. Population Living in Suburbs and Central Cities, 1900–2000
Source: U.S. Bureau of the Census.

central city populations; the comparable figures for the suburbs are 8 percent and 11 percent. However, demographic patterns are also beginning to change in the suburbs, where the proportion of whites declined from 87 percent in 1980 to 75 percent in 2000.

Figure 2.5 shows the extent to which there is a gap in the earnings of persons living in the central cities compared to the suburbs. While city and suburban income gaps appear to be growing in the Northeast and Midwest, they are narrowing in the South and West. Furthermore, as shown in figure 2.6, central cities continue to have higher poverty rates— around 18 percent versus 8 percent in the suburbs.

Most significant, in most of the nation's hundred largest cities, non-Hispanic whites are now in the minority. As shown in figure 2.7, between

Table 2.5. Cities That Grew by More than 30 Percent of Their Populations between 1990 and 2000

City	Population 1990	Population 2000	Percentage Increase
Las Vegas, Nev.	258,296	478,434	85.2
Plano, Tex.	128,743	222,030	72.5
Scottsdale, Ariz.	130,069	202,705	55.8
Boise City, Idaho	125,738	185,787	47.8
Glendale, Ariz.	148,134	218,812	47.7
Laredo, Tex.	122,899	176,576	43.7
Bakersfield, Calif.	174,820	247,057	41.3
Austin, Tex.	465,622	656,562	41.0
Salinas, Calif.	108,777	151,060	38.9
Mesa, Ariz.	288,091	396,375	37.6
Durham, N.C.	136,611	187,035	36.9
Charlotte, N.C.	395,934	540,828	36.6
Santa Clara, Calif.	110,642	151,088	36.6
Reno, Nev.	133,850	180,480	34.8
Phoenix, Ariz.	983,403	1,321,045	34.3
Overland Park, Kans.	111,790	149,080	33.4
Raleigh, N.C.	207,951	276,093	32.8
Chesapeake, Va.	151,976	199,184	31.1
Santa Rosa, Calif.	113,313	147,595	30.3

Source: U.S. Bureau of the Census.

Table 2.6. Cities That Lost More than 5 Percent of Their Populations between 1990 and 2000

City	Population 1990	Population 2000	Percentage Decrease
Cleveland, Ohio	505,616	478,403	−5.4
Washington, D.C.	606,900	572,059	−5.7
Toledo, Ohio	332,943	313,619	−5.8
Jackson, Miss.	196,637	184,256	−6.3
Lansing, Mich.	127,321	119,128	−6.4
Detroit, Mich.	1,027,974	951,270	−7.5
Birmingham, Ala.	265,968	242,820	−8.7
Dayton, Ohio	182,044	166,179	−8.7
Macon, Ga.	106,612	97,255	−8.8
Cincinnati, Ohio	364,040	331,285	−9.0
Pittsburgh, Pa.	369,879	334,563	−9.5
Syracuse, N.Y.	163,860	147,306	−10.1
Norfolk, Va.	261,229	234,403	−10.3
Buffalo, N.Y.	328,123	292,648	−10.8
Flint, Mich.	140,761	124,943	−11.2
Baltimore, Md.	736,014	651,154	−13.0

Source: U.S. Bureau of the Census.

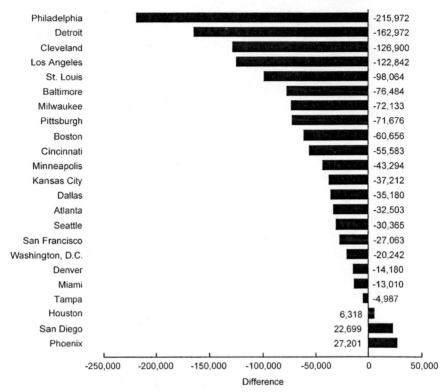

City	Difference
Philadelphia	-215,972
Detroit	-162,972
Cleveland	-126,900
Los Angeles	-122,842
St. Louis	-98,064
Baltimore	-76,484
Milwaukee	-72,133
Pittsburgh	-71,676
Boston	-60,656
Cincinnati	-55,583
Minneapolis	-43,294
Kansas City	-37,212
Dallas	-35,180
Atlanta	-32,503
Seattle	-30,365
San Francisco	-27,063
Washington, D.C.	-20,242
Denver	-14,180
Miami	-13,010
Tampa	-4,987
Houston	6,318
San Diego	22,699
Phoenix	27,201

Figure 2.2.   Central City Employment Change in Manufacturing Trade, 1967–1997
Source: U.S. Bureau of the Census, *City and County Data Book* (1974, 1998).

1990 and 2000, the non-Hispanic white population in those cities fell from 52 to 44 percent—a loss of about 2 million people.[12] The five largest cities alone—New York, Los Angeles, Chicago, Philadelphia, and Houston—lost nearly 1 million non-Hispanic white residents, and the Hispanic populations in these cities grew dramatically, from 17.2 percent in 1990 to 22.5 percent in 2000, representing approximately 3.8 million new Hispanic residents. Ten Texas cities together gained about 1 million Hispanics. Such cities, along with Miami, San Diego, and Los Angeles, have what John Logan and Harvey Molotch refer to as "third world" border city attributes.[13] That is to say, because of the large numbers of low-paid foreign workers who either migrate for the day or week or become permanent residents, these cities provide cheap labor for sweatshop manufacturing or as workers in tourist-related jobs—busboys, dishwashers, or maids. Even in the high-tech field, many small firms cannot afford to go abroad to utilize such persons as low-wage assemblers.

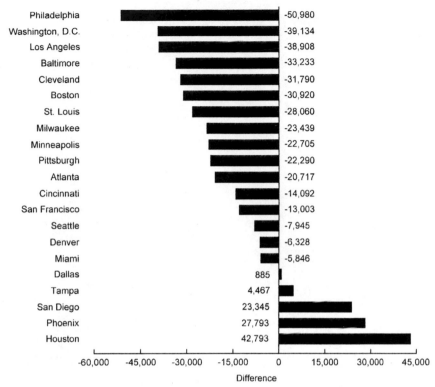

Figure 2.3. Central City Employment Change in Retail Trade, 1967–1997
Source: U.S. Bureau of the Census, *City and County Data Book* (1974, 1998).

Urban Sprawl

As population and development continue to diffuse into the hinterland, huge chunks of open space (rural land) are being developed. In the 1990s, land consumption in the United States proceeded at twice the rate of population growth. As reported by the U.S. Department of Housing and Urban Development, land development between 1994 and 1997 averaged 2.3 million acres annually. Of the more than 9 million acres developed during those years, the overwhelming majority were outside metropolitan areas, in fringe suburbs or smaller towns or cities.[14]

The resulting type of development—commonly known as "sprawl"—is characterized by low density, wide gaps between development clusters, fragmented open space, and the absence of public spaces or community centers. Among the factors that encourage sprawl are heavy public investments in roads, water, and sewer lines; land-use regulations that promote low density and far-flung development; consumers' desire for rural

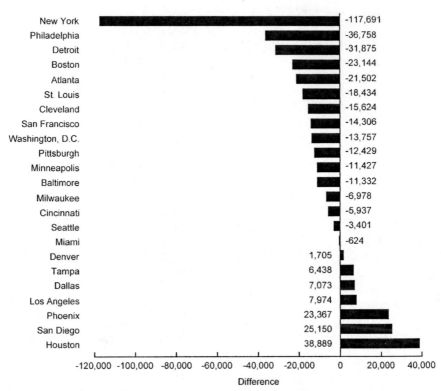

Figure 2.4.   Central City Employment Change in Wholesale Trade, 1967–1997
Source: U.S. Bureau of the Census, *City and County Data Book* (1974, 1998).

Table 2.7.   Distribution of Whites, Blacks, and Hispanics in Central Cities and Suburbs

Race	Year	All Metro Areas	All Central Cities	Suburbs
White, non-Hispanic	1980	77.8	65.2	86.6
	1990	73.3	59.6	82.2
	2000	66.0	51.4	74.9
Black, non-Hispanic	1980	12.4	21.4	6.1
	1990	12.6	21.4	6.9
	2000	12.9	21.1	8.0
Other races, non-Hispanic	1980	2.4	3.0	2.0
	1990	3.9	4.9	3.3
	2000	6.8	8.3	6.0
Hispanic (all races)	1980	7.4	10.4	5.3
	1990	10.1	14.0	7.6
	2000	14.2	19.3	11.2

Source: U.S. Bureau of the Census.

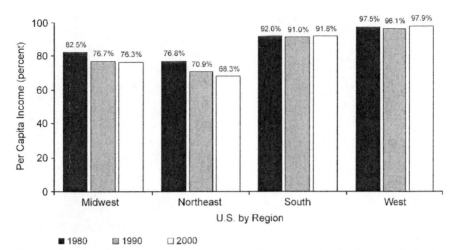

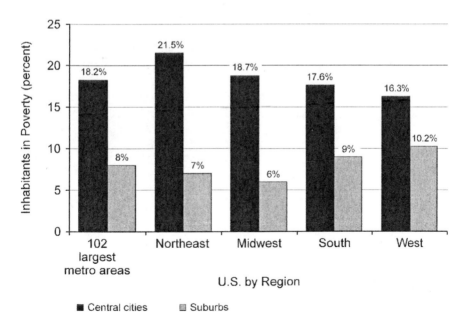

Figure 2.5.   Central City per Capita Income as a Percentage of Suburban per Capita Income by Region: 1980, 1990, 2000
Source: U.S. Bureau of the Census. Reported by Todd Swanstorm, Colleen Casey, Robert Flack, and Peter Dreier, "Pulling Apart: Economic Segregation among Suburbs and Central Cities in Major Metropolitan Areas," Brookings Institution, October 2004, 4.

Figure 2.6.   Percentage of Inhabitants in Poverty, 2000.
Source: U.S. Bureau of the Census.

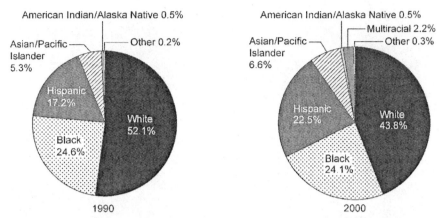

Figure 2.7. Shifts in the Racial and Ethnic Mix in the Top 100 U.S. Cities, 1990–2000
Source: U.S. Bureau of the Census.

lifestyles with large homes and large yards; the preference of business and industry for easy highway access and plenty of free parking; lower land prices and lower taxes in peripheral areas; and advances in communications technology that reduce the need for physical proximity in the course of carrying out various tasks.

Table 2.8 lists the ten metropolitan areas (entities defined by the Census Bureau as central cities and the contiguous development of their suburbs) that did away with the most rural land during the period 1970–1990. Paving and building over hundreds of square miles of woods, wetlands, prairies, and fields, they earned the designation of the nation's "Top Sprawlers."

While various organizations and media persons defend the disappearance of open land as a sign of economic vitality to be encouraged, many

Table 2.8. Metropolitan Areas with Greatest Sprawl, 1970–1990

Rank	Metropolitan Area	Sprawl (sq. miles)
1	Atlanta, Ga.	701.7
2	Houston, Tex.	638.7
3	New York City, N.Y.–N.J.	541.3
4	Washington, D.C.–Md.–Va.	450.1
5	Philadelphia, Pa.–N.J.	412.4
6	Los Angeles, Calif.	393.8
7	Dallas–Fort Worth, Tex.	372.4
8	Tampa–St. Petersburg–Clearwater, Fla.	358.7
9	Phoenix, Ariz.	353.6
10	Minneapolis–Saint Paul, Minn.	341.6

Source: U.S. Bureau of the Census.

others who are concerned with the quality of life contend that sprawl has numerous negative effects—from traffic congestion to air and water pollution to the loss of productive farmland. Sprawl also creates significant economic costs, and one of the most serious of these is the unremitting increase in the demand for roads, utility lines, and service delivery to be extended to dispersed development. Such demands undermine local governments' ability to finance public services. Another economic effect is premature disinvestment in the buildings, facilities, and services of older urban centers.[15]

As long as the U.S. population continues to increase, the expansion of development into rural areas can be expected to continue as well. In light of the growing public awareness of the costs of sprawl, conservationists and slow-growth advocates are beginning to campaign for a smart-growth agenda at all levels of government. For the most part, states have become the focus of a smart-growth policy framework, which includes land-use reforms designated to manage growth at the metropolitan fringe and the use of state resources to preserve tracts of land threatened by sprawl. In addition, states have begun to steer infrastructure investment to older, built-up areas.[16] For example, under Maryland's Smart Growth Act of 1996, state permission or funding for infrastructure and other facilities for new development projects is tightly restricted to older established areas and excludes less-developed outlying areas. Similarly, Oregon has defined an urban growth boundary around the Portland metropolitan area beyond which growth is severely restricted.

From Metropolis to Megalopolis

Spurred by advances in communication and transportation technology, the changing structure of regional economies helps explain how metropolitan areas are being transformed from the monocentric regional model, in which the core city dominates, to a polycentric model based on suburban clustering to most recently a network pattern (see figure 2.8). As described by Allan D. Wallis, the conversion of what were essentially manufacturing-based economies to service-based economies has contributed to new and more complex interdependencies among urban centers within metropolitan regions.

> On the one hand, the concentration of highly specialized services in central cities—especially in finance and law—supports critical control functions needed by modern corporations. On the other hand, the types of innovative/ entrepreneurial activities that are essential to the [functioning] of regions no longer are occurring primarily in central cities, as amply illustrated in the role being played by high-tech suburban corridors like Silicon Valley. But it

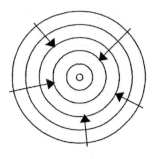

Concentric ring pattern

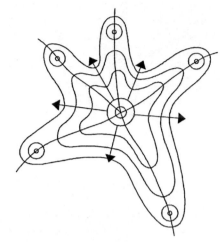

Multiple-nucleus (polycentric) pattern

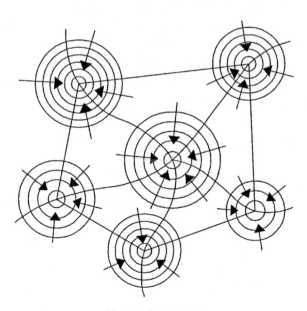

Network pattern

Figure 2.8.   Evolving Patterns of Metropolitan Development
Source: Allan D. Wallis, "Evolving Structures and Challenges of Metropolitan Regions," *National Civic Review* 83 (Winter–Spring 1994): 45.

is not just central cities and suburban municipalities that are drawn into new interdependencies. Suburbs depend on each other for collective economic competitiveness.[17]

These new network patterns and the spread of industry and population have placed increasing importance on the metropolitan area as a geographic unit of analysis. In contrast to earlier times when formal legal boundaries defined cities as incorporated entities, in contemporary times the outer limits of metropolitan regions bear greater significance, for the typical central city cannot be fully understood apart from the suburban neighbors to which it is organically linked. Cities like New York, Los Angeles, Chicago, and Boston lend their names to an entire region to which they contribute only a portion of the population and jobs (table 2.9). Usually surrounding the urban core is a wide array of other public jurisdictions of varying sizes, from the simple to the complex, that collectively comprise the suburban hinterland.

Truman A. Hartshorn and Peter O. Muller have observed that in some metropolitan regions, certain strategically located commercial centers have taken on the characteristics of suburban downtowns, thereby comprising a new type of city.[18] Manifesting a rich mix of diverse facilities, these concentrations typically consist of large shopping malls surrounded by high-tech firms and office buildings that house specialized legal, financial, and retail services. The hinterland of these centers could stretch fifty to one hundred miles. Prime examples are the growth corridors around Route 128 near Boston, which first gave rise to applied high technology; the Schaumburg area near O'Hare Airport, where Sears Roebuck moved its corporate headquarters; and the Perimeter Center off Atlanta's beltway, which is bigger than downtown Atlanta (see box 2.5).

As metropolitan areas continue to expand, they have become linked into a vast urbanized complex that has been dubbed a "megalopolis." French geographer Jean Gottman first discerned this phenomenon in the 1950s as it was emerging along the northeastern seaboard of the United States from Boston to Northern Virginia.

> In this area . . . we must abandon the idea of the city as a tightly settled and organized unit in which people, activities, and riches are crowded into a very small area clearly separated from its nonurban surroundings. Every city in this region spreads out far and wide around its original nucleus; it grows amidst an irregularly colloidal mixture of rural and suburban landscapes; it melts on broad fronts with other mixtures, of somewhat similar though different texture, belonging to the suburban neighborhoods of other cities.[19]

Popularly referred to as Boswash, the region described by Gottman reaches from Boston to Northern Virginia, encompassing such other big

Table 2.9. Metropolitan Areas Ranked by Population, 2000

Rank	Metro Area	Area Covered	Population	Growth*
1	New York	Includes Long Island and parts of N.J., Conn., and Pa.	21,199,865	8.4%
2	Los Angeles	Includes Riverside and Orange counties, Calif.	16,373,646	12.7
3	Chicago	Includes Gary, Ind., and Kenosha, Wis.	9,157,540	11.1
4	Washington-Baltimore	Includes parts of Md., Va., and W.Va.	7,608,070	13.1
5	San Francisco	Includes Oakland and San Jose, Calif.	7,039,362	12.6
6	Philadelphia	Includes Wilmington, Del.; Atlantic City, N.J.; and additional parts of Pa., N.J., Del., and Md.	6,188,463	5.0
7	Boston	Includes Worcester and Lawrence, Mass., and parts of N.H., Maine, and Conn.	5,819,100	6.7
8	Detroit	Includes Ann Arbor and Flint, Mich.	5,456,428	5.2
9	Dallas–Fort Worth		5,221,801	29.3
10	Houston	Includes Galveston and Brazoria, Texas	4,669,571	25.2

*Growth rate from 1990 to 2000.
Source: U.S. Bureau of the Census.

Box 2.5. The Features of the New City

Familiar as we are with the features of the new city, most of us do not recognize how radically it departs from the cities of old. The most obvious difference is scale. The basic unit of the new city is not the street measured in blocks but the "growth corridor" stretching 50 to 100 miles. Where the leading metropolises of the early 20th century—New York, London, or Berlin—covered perhaps 100 square miles, the new city routinely encompasses two to three thousand square miles. Within such "urban regions," each element is correspondingly enlarged. "Planned unit developments" of cluster-housing are as large as townships; office parks are set amid hundreds of acres of landscaped grounds; and malls dwarf some of the downtowns they have replaced.

These massive units, moreover, are arrayed along the beltways and "growth corridors" in seemingly random order, without the strict distinctions between residential, commercial, and industrial zones that shaped the old city. A subdivision of $300,000 single-family houses outside Denver may sit next to a telecommunications research and production complex, and a new mall filled with boutiques once found only on the great shopping streets of Europe may—and indeed does—rise amid Midwestern corn fields.

The new city, furthermore, lacks what gave shape and meaning to every urban form of the past: a dominant single core and definable boundaries. At most, it contains a multitude of partial centers, or "edge cities," more-or-less unified clusters of malls, office developments, and entertainment complexes that rise where major highways converge. As *Washington Post* writer Joel Garreau has observed, Tysons Corner, perhaps the largest American edge city, boasts more office space than downtown Miami, yet it remains only one of 13 edge cities—including Rockville-Gaithersburg, Maryland, and Rosslyn-Ballston, Virginia—in the Washington, D.C., region.

Source: Reprinted by permission of the author, from Robert Fishman, "America's New City: Megalopolis Unbound," *Wilson Quarterly* (Winter 1990): 23.

---

cities as New York, Philadelphia, Baltimore, and Washington, D.C., which serve as more specialized economic subcenters. The number of people who live in this huge geographical cluster approximates 50 million. Other developing megalopolitan centers include the Florida peninsula, the Lower Great Lakes, Northern California, and Southern California.

Global Restructuring

In postindustrial society, as we have noted, the geographic location of labor and capital has come to mean less as firms and industries, scattered at different sites over vast distances, use information technology to coordinate the production and distribution of goods. The big change now taking

place is that this is happening on a global level, involving cities and regions far apart in distant lands. Fiber optics, satellites, and "smart buildings," which are discussed in the next chapter, constantly increase the
speed and efficiency of transactions. Money now moves easily across national boundaries, and stocks, currency, and bonds traded on worldwide
electronic markets amount to an estimated four trillion dollars each day—
nearly twice the annual U.S. budget. International air transport, furthermore, can move products to any continent in one or two days.

Consequently, metropolitan regions have emerged as major centers of
economic activity that participate and compete not only in the national
economy but in the world economy as well. With the ending of the Cold
War and the breakdown of trade barriers between nations, the focus of
competition on the global level has been shifting from the military power
of nation-states to economic entrepreneurship generated on the local-re-
gional level. In light of this, America's economy has come to consist of
constellations of regional economies that are, in effect, replacing the nation-state as key performers in the conduct of global affairs.[20]

Taking note of this, a question of some interest is how America's urban
centers are ranked in the global framework. Saskia Sassen first identified
what she viewed as new hierarchies of power and entrepreneurship with
New York, London, and Tokyo serving as core cities where the world's information and financial superhighway is located.[21] According to her,
other regional and local centers may be considered as being in a second or
third tier to the extent they are able to intersect the global system.

Referring to this, Mark Abrahamson examines the global rankings of
city/regions by identifying several sets of indicators: namely, stock ex-
changes, banking and finance, multinational corporations and foreign direct investment, and corporate services such as advertising, accounting,
and law.[22] According to Abrahamson, the locations of these firms and activities help to define the most important centers of the global economy.
From this, Abrahamson comes up with a composite index as a measure of
that city/region's place in the global hierarchy (see table 2.10). New York,
with a score of 40, the only region to receive the maximum score on every
indicator, comes out on top. London, Paris, and Tokyo are close behind,
with slightly lower scores of between 34 and 37; all four city/regions can
be viewed as being in the first rank. Other North American urban locations that placed in the second and third tiers are Chicago, Los Angeles,
Toronto, San Francisco, and Montreal.

City/regions should be seen as comprising a global network within
which funds flow, information is exchanged, commodities are traded, and
resources are allocated.[23] As places of global accumulation, they accrue
economic and political power over those urban regions lower down in the
hierarchy. Urban centers that cannot compete tend to be outside the global

Table 2.10.　The Global Economic Hierarchy

Score	City/Region
40	New York
34–37	London, Paris, Tokyo
28	Frankfurt
15–16	Chicago, Hong Kong, Osaka, Zurich
11–12	Los Angeles, Milan, Singapore, Toronto
7–8	Beijing, Munich, San Francisco
4–5	Amsterdam, Düsseldorf, Montreal, Seoul, Stockholm, Stuttgart, Taipei

Source: Mark Abrahamson, *Global Cities* (New York: Oxford University Press, 2004), 89.

circuit and are likely to experience increasing levels of unemployment and poverty. Old industrial cities, as we noted, are especially vulnerable.

Of additional significance is that the positioning of city/regions within the global system influences how nation-states, of which they are a part, are able to compete in the global marketplace. The dominant role played by metropolitan areas in the United States has been demonstrated by data compiled by the U.S. Conference of Mayors. The figures show that the value of goods and services produced in the ten metropolitan areas with the largest output in 2000 accounted for about a third of the nation's total. The top three metropolitan areas—New York, Los Angeles, and Chicago—would, if they were nations, rank among the top twenty nations in the world based on the size of their economies.[24]

SUMMARY AND CONCLUSIONS

This chapter has examined how American cities have developed in relation to changes in technology. Table 2.11 reviews the chief characteristics of cities as they have evolved over three overlapping stages: from an earlier agricultural period to a period of manufacturing and heavy industry and, more recently, to an emerging information-/knowledge-based stage. Constrained by limited energy sources such as wind, water, animate power, and the slow speeds of available transportation—walking, horseback, or horse-drawn carriages—most cities in seventeenth- and eighteenth-century America were small and grew very slowly. By the nineteenth century, the advancement of rail transportation contributed to the more rapid growth of cities along the East Coast and opened the West to development in the form of railroad towns. Somewhat later, electric trolleys made it possible for people to live further from their places of work and still get to their jobs on time. When automobiles became available in the 1920s, new suburban communities developed as commuting zones for the central

Table 2.11. Key Characteristics of Cities in the Preindustrial, Industrial, and Postindustrial Stages

Characteristic	Preindustrial	Industrial	Postindustrial
Time frame	17th and 18th centuries	1830s and later	1950s and later
Power sources	Wind, water, animate	Steam/internal combustion	Superconductivity
Technological advancement	Agricultural tools	Energy	Information
Role of technology	Extraction	Fabrication	Process
Main product	Food	Commodities	Knowledge
Social institution	Family farm	Mills, factories	R&D centers
Main labor force	Farmers	Factory workers	Information
Main mode of movement	Pedestrian, horse and carriage	Rail, auto	Telecommuting
Division of labor (skills)	Simple	Highly specialized (routine)	Very highly specialized (customized)
Division of labor (geographical)	City	Regional	International
Marketplace	Commons	Central business district/mall	Electronic network
Urban pattern	Human network	Monocentric	Global networks
Social pattern	Integrated	Segregated	Highly segregated

Source: Adapted from Tarik A. Fathy, *Telecity: Information Technology and Its Impact on City Form* (New York: Praeger, 1991), 28.

cities. Since the 1950s, advancements in communications and transportation technology have accelerated outward expansion into what used to be called the hinterland, and urban spatial patterns have been undergoing radical change, from a monocentric pattern—where central cities dominated the greater region—to polycentric networks of urban-suburban conglomerations.

Postindustrial urban economies can be outlined schematically. First, business and industry have been changing the ways they conduct their affairs. Greater specialization and customization of services and increased emphasis on research and development and on information processing have combined to replace mills and factories with offices as the primary workplaces. Somewhere between half and two-thirds of all employees work in offices. While the higher levels of management continue to rely on close physical contact and face-to-face interaction as necessary ingredients of decision making, the greater number of office positions are involved in routine paper pushing or computer processing. Persons in these jobs do not need to be in the same location. Utilizing computer technology, more and more workers are being relocated to distant back-office locations where costs are lower and, as we shall see in the next chapter, more and more persons are working at home.

Second, as we have observed, manufacturing in the United States has been declining, from a third of all jobs at the end of World War II to less than a quarter of all jobs in recent times. Of greater significance is that advances in communications mean that those remaining industries do not necessarily have to be located near sources of power and raw materials. Most manufacturers are now more dependent on specialized knowledge than on raw materials such as coal and ore, and industries are now free to follow the workers to wherever they prefer to live.

Third, the dispersion of population within metropolitan areas has been largely white, creating large-scale segregation of poor blacks and Hispanics in older central cities. As metropolitan areas have developed, the jobs best suited to the type of labor force locked into the central cities have moved to the suburbs.[25]

Fourth, automobiles, airplanes, telephones, and, more recently, wireless and computer networks have made most places accessible over ever greater distances. Business and consumption patterns that were previously restricted to a relatively small proportion of the population have now become highly diffused. Though the concept of a marketplace or a central business district persists in many cities, many other urban communities are relying increasingly on electronic media for conducting commerce and trade.

Fifth, different metropolitan communities have come to perform different roles in national and international multicentered urban networks. In these larger systems, roles are not so much territorial as functional in nature. Each metro area provides some services and goods for the others, and each, in turn, is served by others. Together they constitute a complex web of urban interdependencies. As the range and frequency of transactions facilitated by electronic media continues to rise, these urban networks continue to grow, in many instances reaching global proportions. At the same time, those metros that do not fit into the new global economy risk decline.

NOTES

1. Alvin Toffler and Heidi Toffler, *Creating a New Civilization: The Politics of the Third Wave* (Atlanta: Turner, 1994).
2. Carl Bridenbaugh, *Cities in the Wilderness: The First Century of Urban Life in America, 1625–1742* (New York: Ronald Press Co., 1938), 467.
3. Edward Augustus Kendall, *Travels through the Northern Parts of the United States, in the Years 1807 and 1808* (New York: I. Riley, 1809), vol. 3.
4. Sam Bass Warner Jr., *The Private City: Philadelphia in Three Periods of Its Growth* (Philadelphia: University of Pennsylvania Press, 1968).
5. David R. Goldfield and Blaine A. Brownell, *Urban America: A History*, 2nd ed. (Boston: Houghton Mifflin, 1990), 77.
6. John Hoyt Williams, *A Great and Shining Road: The Epic Story of the Transcontinental Railroad* (New York: Times Books, 1988).
7. David E. Nye, *American Technological Sublime* (Cambridge, Mass.: MIT Press, 1994), 120–23.
8. Nye, *American Technological Sublime*, 146–49.
9. Arthur Meier Schlesinger, *The Rise of the City, 1878–1898* (New York: Macmillan, 1933), 79, 80.
10. See Martin Wachs and Margaret Crawford, *The Car and the City: The Automobile, the Built Environment, and Daily Urban Life* (Ann Arbor: University of Michigan Press, 1991).
11. Brian J. L. Berry, *The Human Consequences of Urbanisation: Divergent Paths in the Urban Experience of the Twentieth Century* (New York: St. Martin's, 1973), 48.
12. Brookings Institution, Center for Urban and Metropolitan Policy, *Racial Change in the Nation's Largest Cities: Evidence from the 2000 Census* (Washington, D.C.: Brookings Institution, 2001), available at www.brookings.edu/es/urban/census/citygrowth.htm.
13. John R. Logan and Harvey L. Molotch, *Urban Fortunes: The Political Economy of Place* (Berkeley: University of California Press, 1987), 273.
14. U.S. Department of Housing and Urban Development (HUD), *The State of the Cities 2000* (Washington, D.C.: HUD, 2000), 40, 41.
15. See Gregory D. Squires, ed., *Urban Sprawl* (Washington, D.C.: Urban Institute Press, 2002).
16. Bruce Katz, "Smart Growth: The Future of the American Metropolis," Case Paper 58 (London: Center for Analysis of Social Exclusion, London School of Economics, 2002).

17. Allan D. Wallis, "Evolving Structures and Challenges of Metropolitan Regions," *National Civic Review* 83 (Winter–Spring 1994): 40–53.

18. Truman A. Hartshorn and Peter O. Muller, "Suburban Downtowns and the Transformation of Metropolitan Atlanta's Business Landscape," *Urban Geography* 10 (1989): 375–95.

19. Jean Gottman, *Megalopolis: The Urbanized Northeastern Seaboard of the United States* (New York: Twentieth Century Fund, 1961), 5.

20. Neal R. Peirce, with Curtis W. Johnson and John Stuart Hall, *Citistates: How Urban America Can Prosper in a Competitive World* (Washington, D.C.: Seven Locks Press, 1993); see also Kenichi Ohmae, *The End of the Nation State: The Rise of Regional Economies* (New York: Free Press, 1995).

21. Saskia Sassen, *The Global City: New York, London, Tokyo*, 2nd ed. (Princeton, N.J.: Princeton University Press, 2001).

22. Mark Abrahamson, *Global Cities* (New York: Oxford University Press, 2004).

23. See also Peter J. Taylor and Robert E. Lang, "U.S. Cities in the 'World City Network'" (Washington, D.C.: Brookings Institution, 2005), available at www
.brookings.edu/metro/pubs/20050222_worldcities.pdf.

24. U.S. Conference of Mayors, *U.S. Metro Economies* (Washington, D.C.: Conference of Mayors, 2001).

25. See Michael A. Stoll, "Job Sprawl and the Spatial Mismatch between Blacks and Jobs," (Washington, D.C.: Brookings Institution, 2005), available at www
.brookings.edu/metro/pubs/20050214_jobsprawl.pdf.

3

Third Wave Technologies and Spatial Restructuring

Terms such as *information superhighway*, *digital society*, and *cyberspace* are commonly used to describe the information revolution that is rapidly transforming society into what we call the Third Wave. Broadly speaking, such references pertain to innovations in telecommunications technology that merge television and computers into digital streams of sounds, images, and text that transmit information instantaneously almost anywhere. In less than two decades, the telecommunications industry has changed from a slow-moving, largely ignored sector to the world's largest and fastest growing industry. Through its impact on almost all aspects of people's lives and work, it is significantly reshaping settlement patterns all across the American landscape. To better explain this phenomenon, the present chapter focuses on the various devices and techniques that transmit information via wire, optical fiber, optical wireless, or radio wave.

TELECOMMUNICATIONS SYSTEMS

To begin to comprehend the role of telecommunications in contemporary society, it is necessary to look at its earliest beginnings. As a precursor to today's sophisticated communications machines, the first device to utilize electronic energy for the rapid transfer of information over distance was the telegraph. Invented by Samuel F. B. Morse in 1837, telegraph communication was based on a system of dots and dashes representing letters and numbers. The first demonstration of the system by Morse was conducted for his friends at his workplace. In 1843, Morse obtained financial support from the U.S. government to build a demonstration telegraph system thirty-five miles long between Washington, D.C., and Baltimore. Wires were attached by glass insulators to poles alongside a railroad. When the system was completed, public use was initiated on May 24, 1844, with transmission of the message "What hath God wrought." Subsequently, by 1851 telegraph companies were operating in more than fifty cities in the United States; a few years later, in 1858, the first telegraph cable across the Atlantic entered service, sparking celebrations in cities up and down the Northeast. Thus was inaugurated the telegraph era in the United States, which was to last more than one hundred years.

At the same time that Morse was experimenting with his invention, another inventor by the name of Alexander Graham Bell was working on a method of sending multiple messages on the same telegraph line. Instead of using dots and dashes, he found a way to send the human voice. His first words, "Mr. Watson, come here, I want you," were transmitted a short distance through wire in 1876. Subsequently, Thomas Edison and others made significant improvements in the telephone by increasing the electrical current that carried the audio. By the early 1900s, many households in America and Europe had their own phones, and in 1915 another milestone was achieved when Thomas A. Watson received the first transcontinental phone call from his fellow inventor, Alexander Graham Bell.

These two methods of communication—telegraph and telephone— which have served effectively for transmitting information well into the twentieth century, have important implications for how the civilized world will communicate in the twenty-first century. While both devices rely on changes in electrical current to convey messages, they are different in that the telephone utilizes an analog form for the encoding and decoding of information while the telegraph utilizes a digital form.

Bell devised a method whereby words spoken into a mouthpiece would cause a diaphragm to vibrate in response to the sound waves generated by the person's voice. Bell's equipment could then transform the subtle modulations in pitch into a continuously changing electrical wave pattern that was analogous to the sampled speech. At the receiving end, the electrical wave was decoded into sounds—language—understandable to the listener.

The digital method instead converts sound waves into numerical values of 1s and 0s. In the circuits of a computer's memory, 0s travel as low-voltage electricity, turning off switches called *transistors*; 1s, traveling at a higher voltage, turn on the switches. These units of information are binary digits, or bits. This binary encoding of information underlies all recent advances in information technology. Utilizing digital technology, the computer provides low-cost, high-speed processing and transmission of data,

Box 3.1. A Brief History of Computer Technology

1940s	The first electronic computer is developed. It is called the Electronic Numerical Integrator and Computer (ENIAC) and can calculate with great speed. Bell Labs develops the transistor, which initiates the second generation of computers.
1950s	IBM makes the transition from punch-card calculators to electronic computers with the 701—bought mostly by the military.
1960s	Control Data introduces the 6600, the first real supercomputer, designed by Seymour R. Cray.
1970s	Digital Equipment Corporation dominates the market with the PDP, the first minicomputer that can easily be used by corporate departments.
1980s	IBM introduces the IBM Personal Computer (PC), thereby becoming the leader in the microcomputer market. The Apple Macintosh, which users can control with a mouse, ushers in the era of personal computers.
1990s	PalmPilot's handheld computer promises a shift from desktop to shirt pocket.
2000s	Multitask mobile cell phones are introduced that allow subscribers to take pictures, access the Internet and e-mail, record video clips, make phone calls, engage in wireless gaming, and download TV programs. The rush to turn these devices into all-purpose pocket pods promises to keep coming.

thus transforming conventional methods of communicating in our homes and in our places of business.

Though people have used calculating devices such as the abacus for thousands of years, the first person to come up with the idea of an information processor was Charles Babbage, a nineteenth-century English mathematician. He designed an elaborate calculating machine, which he called an "analytical engine." Although Babbage failed in his attempts to build it, the analytical engine manifested most of the mechanical elements of modern electronic data processing. The first electronic computers appeared in the 1940s (see box 3.1). Their electronic switching circuits were based on vacuum tubes that looked like lightbulbs. Unfortunately, lightbulbs produce a great amount of heat, and they eventually burn out. Employing thousands of vacuum tubes in one electronic computer meant numerous burnouts, which in turn meant substantial loss of time. It also limited the range of problems that computers could solve.

First developed in 1948 by Bell Labs, transistor circuits soon replaced vacuum tubes in computer applications. About the size of a dime, transistors worked much faster than tubes and eventually became much less expensive; they also produced less heat, and they rarely failed. Subsequently, multiple transistor circuits were combined onto a single chip, creating an integrated circuit. Computer chips now in use are integrated circuits containing the equivalent of millions of transistors packed into less than a square inch of silicon. Although they are paper thin, they are really three-dimensional packages. Microscopically thin layers are subdivided by function. Each layer is integrated with the layers above and below, like floors in a building linked by wiring and plumbing. The speed and power of today's—and tomorrow's—electronics depend on how many components, such as circuits and transistors, engineers can squeeze onto the tiny, programmable silicon chips that are the brains of computers, video games, pocket cellular phones, and other devices. As the power of chips continues to grow, computers become smaller, faster, less costly, more capable, and more reliable. Engineers believe that in the future, it will be possible to build chips that can store and retrieve ever greater amounts of memory.

In his book, *The Road Ahead*, Microsoft founder Bill Gates states that to understand why information is becoming so dominant in virtually all aspects of society, it is necessary to know how digital technology is changing the way we handle information.[1] The most notable difference is that in the future almost all information will be digital. At the present time, whole print libraries—including newspapers and magazines, photographs, films, and videos—are being converted into digital electronic information, which is stored in computer databases. Digital devices such as the compact disc player, the cashier scanner, and the computer function on the principle of converting data into binary language. As a consequence, the way in which we conceptualize information is also undergoing change.

During the industrial age, information was knowledge. To gain understanding of any particular subject, the individual would conduct interviews, search for relevant documents, and collect data. Through such research, knowledge could be increased. Of course, this definition remains applicable. During World War II, however, the term *information* began to be redefined. Technicians developed methods for sending coded signals over busy radio channels, and antiaircraft guns were designed to respond automatically to data being fed in. Martin Gay explains, "In a real sense, the way humans interacted with information began to change; it was translated into a commodity that could be used to operate a machine, guide a missile, and finally drive an entire economy."[2]

Though we can anticipate further rapid development of telecommunications in the twenty-first century, some difficult questions remain to be resolved. Will the stream of data that connects firms and households travel on high-frequency wireless waves through the air, like radio and television signals? Will it ricochet off satellites in space? Will it be carried over cable or telephone wires? Most likely, all these techniques will be used. For the present, the widespread presence of telephones in nearly every household and business makes the phone the immediate choice for carrying data almost anyplace in the industrialized world where there is a computer and a modem. Certain constraints must be overcome, however, before traffic can flow freely on the information highway.

At present, most consumers who connect into online services such as the Internet use the telephone system's conventional twisted-pair copper wiring, a narrowband mechanism that utilizes analog tones to communicate data. Modems (shorthand for modulator-demodulators) convert a computer's digital information into tone patterns that telephone networks can carry, and vice versa. Unfortunately, there is a limit to the speed at which such data can be transmitted. Consequently, telephone and cable companies are upgrading their systems with new fiber-optic cable. Composed of extremely fine glass threads, such cable has considerably greater bandwidth than copper wire. Such systems can connect people to interactive gaming, banking, movies, medicine, buying, selling, reading, publishing, registering, voting, learning, teaching, and all forms of human transactions.

Until 1996, a broadband interactive network that carried both video and telephone services was against the law in the United States. Burdensome regulations restrained phone companies from utilizing video and prevented cable companies from providing telephone services. The Telecommunications Act of 1996 was designed to eliminate many barriers to competition among broadcast media, long-distance providers, local phone companies, the cellular industry, and cable. As an important side benefit, they have been able to build midband and ultimately broadband networks that carry entertainment as well as communications.

WIRELESS COMMUNICATIONS

Wireless communication has been in existence nearly as long as wired communication. In 1896, only twenty years after Alexander Graham Bell invented the telephone, Guglielmo Marconi demonstrated to the British Post Office a new system of point-to-point communication. Called "wireless telegraphy," Marconi's invention did not carry voices. It worked by producing radio frequency currents in an antenna, whose spark then transmitted emissions to another antenna at some distant location. By giving "on" and "off" signals, it acted as a kind of primitive digital system. Though very limited in range and slow in sending and receiving signals, spark transmission was first used by the British Navy for short-range, ship-to-ship communication.

It wasn't until early in the twentieth century that engineers developed technology that could produce and receive tuned, continuous-wave transmissions. The first transmitters were mechanical instruments based on fast-revolving electrical alternators. In the United States, the RCA Corporation was formed with government assistance to protect U.S. interests in what was perceived as a form of military technology. The nation had not yet foreseen the eventual rise of radio broadcasting and how it would transform American culture and lifestyles.

Though vastly improved over the years, the basic principle of wireless communication as first developed by Marconi still holds. It is the ability to generate electromagnetic waves that makes possible the transmission of information over distance without wired connections. The frequency of an electromagnetic or radio wave is its oscillation rate measured in cycles per second, or hertz (Hz). The range of radio frequencies that facilitate communications begins at a few thousand hertz and rises up to a few hundred billion hertz. Radio systems range from the low-cost cordless telephone with a very low-power transmitter and a simple antenna, to high-power transmitters carrying multiple information signals and using complex directive antennas. The distance covered may range from a few feet, in the case of a cordless telephone, to millions of miles in the case of space satellites.

There is a surprisingly wide array of wireless services now offered or under development. Their special advantage is that they allow communication for people on the move. Using a common handset and a single telephone number, customers no longer need to be associated with a physical location, and calls can be routed to where the individual happens to be.

At present, there are four types of mobile radio service.[3] One type consists of one-way paging, so the user (e.g., a plumber or a medical doctor) carries a small device called a "beeper" to receive one-way messages. A second type is two-way dispatch, as used by taxis, police, firefighters, and tow trucks. It is typically a command-and-control system, which requires coordination among units. A third type is the two-way mobile/portable telephone as represented by cellular and other personal wireless telephone services. It allows the user to place and receive ordinary telephone calls on a one-tone basis. The fourth basic mobile radio service is two-way data communications, a newly emerging service that has come to dominate wireless services. It allows various forms of wireless data communications, such as computer-aided dispatch, electronic messaging/mail, and computer-to-computer communications. In the transportation and logistics industries, for example, wireless technology is used for fleet management and tracking the delivery of goods. Wireless satellite-based tracking systems enable trucking companies to know the exact location of each truck in their fleets.

In the 1990s, the federal government decided to boost competition in telecommunications services by auctioning different spectrums of the public airwaves to analog cellular and digital phone providers. By 2000, according to the Federal Communications Commission (FCC), more than a hundred thousand transmission sites covered the nation. Though the construction of transmission towers often incites zoning battles, pitting neighborhood residents against wireless networks phone companies, Congress and the FCC argue that the budding wireless networks will potentially benefit everyone through price-cutting competition with traditional wire-line phone companies.

By the dawn of the new century, electronic communications, once thought to be permanently bound to the world of cables and hard-wired connections, suddenly were sprung free. Mobile cells phones were being adopted by consumers at an accelerating rate. Whereas in 1984 only 40,000 people used such devices, by 2003 approximately 180 million people were using them and approximately 63 percent of all Americans owned cell phones. Payphones, on the other hand, showed a moderate decline from 2.1 million in 1997 to 1.5 million in 2003.

Not too long ago, mobile phones were unwieldy and expensive novelties. Today they are pocket-size powerhouses that subscribers use to take pictures, access the Internet and e-mail, and record video clips, in addition to making phone calls. The push by wireless carriers and tech companies to turn these devices into all-purpose pods will continue as flashy new features are rolled out, such as using them to track your children's location (see box 3.2), for wireless gaming, for downloading TV programs, as credit cards, and to give audio directions on how to get from point A to point B. It should be noted, moreover, that while this device allows users to call anywhere anytime, it also obligates them by allowing them to be called anytime, anywhere.[4]

THE INTERNET AND THE
WORLDWIDE INFORMATION EXPLOSION

Many of the early applications of information technology were designed primarily to improve the internal operations of organizations. Consequently, they often created "islands of automation" with little interconnectedness between component units.

Computer networks did not exist anywhere. An important breakthrough occurred in the late 1960s during the Cold War when the Department of Defense (DOD) undertook the development of a fail-safe communications mechanism that could initiate a military response after a nuclear attack. The DOD's Advanced Research Projects Agency (ARPA)

Box 3.2. Should Students Have Cell Phones in School?

As safety issues come to assume greater urgency in the schools, parents have been asserting pressure on school administrators to relax rules on student use of cell phones, for such devices have become tools by which parents can keep in touch with their children during the school day. As one parent exclaimed, "It's for my child's security and my peace of mind."

It is estimated that, as of 2004, half of American teens from thirteen to seventeen years of age had a mobile phone; in many high schools, administrators estimate the figure as being closer to 90 percent. Though most schools restrict the use of phones during class time as being disruptive, more than half a dozen states, including Georgia, Louisiana, Michigan, Maryland, and Nevada, and numerous school districts have relaxed their bans or abolished them entirely. New policies vary widely. While some schools permit phones to be displayed in plain view, others require the phones to be stowed in purses, pants, or backpacks. In some cases, certain types of phones, such as those with cameras, remain banned.

Source: Matt Richtel, "School Cellphone Bans Topple," *New York Times*, 29 September 2004.

---

funded a project to create computer communications among its university-based researchers. With ARPA's backing, a small group of researchers and engineers began a quest to connect computers across the country.[5]

In 1969, ARPA awarded a contract to Bolt, Beranek, and Newman, a small Cambridge, Massachusetts, company, to build a computerized switch called the interface message processor (IMP). Nine months later, the first IMP was installed at the University of California, Los Angeles, and three other university sites. Soon, a national network called ARPANET evolved from those four sites. Protocols were designed, and along the way a series of accidental discoveries were made—one of which was electronic mail. Right away, "e-mail" became the most popular feature of the Net. As ARPANET continued to grow, it merged with other computer works to become what is presently called the Internet. ARPANET itself was phased out in 1990 when some 313,000 hosts were online all over the world. At this time, personal computers (PCs) were becoming a more common appliance for access to the Internet, and by 1992 the number of hosts had reached more than a million.

Around 1989, Tim Berners-Lee of the European Laboratory for Particle Physics in Switzerland had proposed a new Internet protocol that he called the World Wide Web (WWW, or simply the "Web"). His new software was designed to link information in various databases around the world. By the summer of 1991, Berners-Lee was sending out messages to

Table 3.1.   Internet Usage Statistics—The Big Picture

World Region	Internet Usage	Usage Growth, 2000–2005	Penetration (% of population)
Africa	13,468,600	198.3%	1.5%
Asia	302,257,003	164.4	8.4
Europe	259,653,144	151.9	35.5
Middle East	19,370,700	266.5	7.5
North America	221,437,647	104.9	67.4
Latin America	56,224,957	211.2	10.3
Oceana/Australia	16,269,080	113.5	48.6
World Total	888,681,131	146.2	13.9

Source: "Internet Usage Statistics—The Big Picture," InternetWorldStats.com, 2006, www.internetworldstats .com/stats.htm.

various Internet locations, inviting people to link up to his lab and download the basic software needed to create Web documents.

In its early form, the Web was not graphical but was based instead on text and typed commands. Marc Andreessen and a small group of graduate students at the University of Illinois developed Berners-Lee's free software into Mosaic, a program that would run on a user's computer and turn the Web into a graphical wonderland. Now it was possible to move around the Web just by using a computer's pointing device to click on words and symbols on an attractive screen. Subsequently, Andreessen cofounded Netscape Communications Corporation, which became the world's leading Web browser.

By 1996, more than five million hosts were sending information to the Internet. Estimates at that time put the total number of Internet users at thirty million, a figure that doubled in less than a year. A goodly number of people who receive information on the Web give some back in the form of electronic "home pages." These are collections of menus that can include text, colorful pictures, and short video and sound clips.

The Web has now become the center of action in the development of the information superhighway. It is estimated that at the time of this writing, North America alone had more than 221 million Internet users—in excess of two-thirds of the continent's population (see table 3.1). This represents a usage growth of 105 percent from 2000 to 2005. Table 3.1 shows that while North America continues to lead in the use of Internet technology, the rest of the world is beginning to catch up.

In one sense, the Web is just another communications tool, accessible to anyone with the proper equipment—a computer, a telephone line, a subscription to an access service. In another sense, it advances far beyond anything ever produced in the history of information technology and has the potential to radically change how people interact. For one thing, it provides the means for any person to transmit a message of any length or content to millions of other persons. Second, it allows people to communicate effortlessly between one location on the globe and another. Third, it can bring into view entire stores of information—in books, film, lectures, articles—making them available anywhere at any time.

What's more, for better or worse, people have to come to rely on their own efforts in meeting needs. For example, in the past, an individual planning a trip would refer to an authoritative source such as a travel agency. These days, however, many travelers "surf" the Web and find sites overflowing with pictures and helpful advice created by vendors, government tourist agencies, recent travelers, and others. More and more people book their own flights and make their own travel arrangements through the Internet. As another example, the practice of medicine is also being affected. Instead of consulting with doctors, people with medical problems are using the Web to find out how they should treat their ailments. While some people enjoy this newfound freedom to access ideas, the downside is that not all voices are equally credible and that it can be risky to ignore professional advice.

To convey a broader understanding of how the Internet, cell phones, and other electronic devices are affecting the way we do things, a wide-ranging discussion of different applications and their implications follows in the next section.

ELECTRONIC APPLICATIONS

The full range of electronic applications is inestimable. Increasingly, electronic linkage is substituting for physical accessibility, and storage of bits is displacing storage of physical artifacts such as books and documents.[6] Consider the typical bookstore. The problem with printed books and magazines is that while they can be mass-produced at a central location, they must then be stored in a warehouse, transported to retail outlets, and subsequently hand-carried and read by interested consumers. Each of these activities requires a specially designated facility such as a printing plant, a warehouse, a bookstore, and ultimately a comfortable chair in a well-lighted room. These facilities are distributed at appropriate locations in the urban setting. However, where bookstores can download texts and laser-print them, producers and wholesalers save on inventory, warehouse, and transportation costs; furthermore, customers can select from a wider inventory of reading material. A somewhat different publishing strategy is to download books and documents from online databases to home laser printers and to download recordings at home recorders. When the Internet Company's Electronic Newsstand opened in 1993, it provided online access to magazine articles, thereby allowing customers to browse as they would a typical newsstand.

In a similar fashion, virtual museums and galleries can present popular exhibits free of long lines of viewers moving slowly from one object to the next. Digital images of art, sculptures, and specimens of all kinds can be viewed on a personal computer or in a video theater, making large exhibition spaces unnecessary. Each object on view can have links to other objects that comprise a pattern designed by the viewer according to his or her personal interests.

Presently, libraries connected to the Web function a full twenty-four hours a day. Computers disseminate information to anyone who comes calling anytime during the day or night. Nor does it matter where those libraries are located. It is just as easy for a person in Cape Town, South Africa, or Melbourne, Australia, to link electronically into the Library of Congress as it is for someone in Washington, D.C.

Even more impressive is an ambitious new plan announced by the search engine Google to start converting millions of books into digital files in collaboration with several major libraries, including the libraries at Harvard, Stanford, and Oxford universities as well as the New York Public Library. The goal is to organize the world's information and make it universally accessible. Google estimates will take six years to scan approximately 15 million books.[7]

In the new millennium, the idea of an electronic college campus that will replace or parallel the physical one seems increasingly plausible. Indeed, educators are beginning to explore the full potential of distance learning through video instruction, thereby precluding the need to assemble students in large lecture halls. Here the classroom is no longer a place for lecturing to relatively small groups of students, but a place for directing streams of information to faraway locations where students participate through online communication. On some college campuses, video conferencing allows students to participate in seminars and consult with faculty from their dormitory rooms. In a similar vein, educators are predicting that in the very near future, school will be vastly different for children than it was for their parents. Sitting at home-learning stations, students will be able to access vast sources of knowledge by engaging in hands-on learning experiences much like video games.[8]

In the medical field, telemedicine is emerging. Net-savvy physicians can now treat patients thousands of miles away by using electronic networks to see their X-rays and medical histories. Doctors and patients don't have to coordinate their schedules, because doctors can use e-mail to answer questions that arrive overnight from a different time zone. As more homes get on the Net, diagnostic devices will facilitate virtual house calls. By utilizing sensors to monitor such symptoms as heart rate, respiration rate, temperature, and blood pressure, remote continuous care could also be provided; in certain rare instances, even surgical procedures could be administered through digital robotics. Perhaps most promising is that isolated and immobilized patients who require medical treatment would be able to avoid difficult and time-consuming travel.

Financial Services

Over the last two decades, banking and related financial services have also moved to cyberspace. Bank buildings, like other institutional structures, are no longer where the money is. Automatic teller machines, first introduced by Citicorp in 1971, now process more than half of all bank transactions. At the same time, banks have been launching marketing campaigns to encourage customers to do most of their banking from home through their personal computers.

Financial institutions have been examining three different ways of linking personal computers to the bank. All three require a modem. One method allows customers to dial directly into the bank's computer over phone lines. Another connects with customers through a commercial on-line service. The third method connects through the Internet. A broad array of financial services is already available from the personal computer keyboard—including loans, life insurance, real estate, and financial planning. About the only financial service that the personal computer cannot do right now is disburse cash—and that's being worked on.

Over time, smart cards may replace cash as the means by which Americans make many of their payments and purchases. Using either digital cellular telephones or a new generation of computers equipped with card readers, individuals may soon be able to download money from their bank accounts directly onto the cards. Unlike traditional credit and debit cards, smart cards have a thumbnail-size computer chip embedded in the plastic. These microprocessors can store thousands of times more information than conventional credit cards backed with magnetic strips. It is anticipated that a single smart card could replace a handful of credit and debit cards, serve as a driver's license, store an individual's medical history, feed a parking meter, and serve as a tamperproof personal ID encoded with a person's fingerprint. When it comes to smart cards, Americans are technological laggards. Today, smart cards are commonly used in Europe—and nearly everywhere in France, where they've replaced coins in everything from telephones to washing machines.

One likely consequence from the growing use of electronic payments is that the check—that piece of paper that has been an integral part of American financial life—is headed toward oblivion. Many consumers and businesses say checks are so antiquated and expensive that their demise can't come soon enough. A new law, the Check Clearing for the 21st Century Act, which passed in 2004, will eliminate many of the check's few remaining advantages for consumers. It will allow banks, retailers, and others to replace the paper checks they receive with electronic versions. The potential savings have been envisioned with glee by most businesses.

Perhaps no service industry has seen so much affected by the information revolution as the financial securities industry. It was once a genteel, paper-based business, but telecommunications technology has made financial markets bigger and faster and in some ways more volatile. Electronic networks now deliver price data, new company earnings, and other information to computer screens everywhere. In handling customer accounts, they permit traders to move money across international borders at the touch of a button. Using desktop computers, traders can zap billions of dollars between markets around the world. Because these markets are electronic, they're all linked together. In effect, there are no more local markets, only global ones.[9]

Overall, local experts contend that technology has had a positive impact. Along with worldwide deregulation, it has created a quick and efficient way to move money from those who have it to those who need it. Stocks, currency, and bonds traded on worldwide electronic markets amount to an estimated four trillion dollars each day, nearly twice the annual U.S. budget. Consequently, banking and finance are now no longer locally oriented activities but important global industries in their own right. Financial corporations the size of Citibank or Salomon Brothers can influence the rate of economic development throughout the world. But while the growth of international finance has helped world trade double in the past score years, creating jobs and improving livelihoods across the globe, it leaves little room for error. The electronic networks can whisk money away just as swiftly as they deliver it, causing economic crisis. Mexico's government found this out in 1994, and Asian countries like Thailand and Indonesia in 1997 and 1998.

E-Commerce

In the area of merchandising, a number of retailers and students of marketing are anticipating an era of teleshopping that will eventually replace conventional shopping. R. Tompkins, for example, believes that most forms of retailing will be situated within virtual electronic space, which will eliminate the physical inconveniences involved in conventional shopping.

> [When] virtual reality enters the world of retailing . . . consumers will be able to try on clothes by watching computer-generated images of themselves (or,

indeed, of someone else) wearing them. Later, instead of watching a video screen, people will probably be able to don a virtual reality helmet and gloves, then transport themselves into the stores of their choice. They will roam the virtual aisles, examining virtual goods and quizzing a virtual sales assistant for more information if required.[10]

Others do not agree with this prognostication. As Judy Hillman explains, "The telestore may be too remote, except for the housebound. . . . For most people, shopping is likely to remain a social, visual, tactile, stimulating if sometimes exhausting, acquisitive experience, which brings particular pleasure in the unexpected bargain or encounter."[11]

Indeed, the history of teleshopping shows a mixed record of hits and misses. In 2004, online retailing accounted for 4.6 percent of total retail sales in the United States, according to data released by the National Retail Federation's Shop.org subsidiary. Approximately 69 percent of households in America that use the Internet make online purchases, according to Forrester Research.[12] Consumer spending on U.S. retail sites has grown from $53 billion in 2001 to $117.2 billion in 2004. eBay has been the top player, with more than 60 million active users ready to swap more than $40 billion in goods and services at the time of writing (see table 3.2.).

Founded in the living room of Pierre Omidyar's home in 1995, eBay was intended from the start to be a marketplace for the sale of goods and services for individuals. What used to be conducted through

Table 3.2.   Top Fifteen Retail Sites

	Unique Visitors in May 2005 (in millions)
eBay	64.3
Amazon	41.0
Wal-Mart	21.3
Shopping.com	19.8
Target	19.5
Apple Computer	17.1
Shopzilla.com	16.2
Overstock.com	16.1
Moviefone	15.3
Cingular.com	14.4
American Greetings Property	14.3
Ticketmaster	13.6
Yahoo Shopping	12.9
Dell	11.6
1-800-Flowers.com	11.5

Sources: comScore Network and eBay, as reported in Leslie Walker, "E-Commerce's Growing Pains," *Washington Post*, 25 June 2005.

garage sales, collectibles shows, and flea markets could now transpire swiftly and efficiently on a global level. Buyers and sellers are brought together in such a way that allows sellers to list items for sale and buyers to browse and bid on items of interest. Browsing and bidding on items is free, but sellers are charged a fee based on the asking price of each item.

Table 3.2 lists other top retail sites. Most prominent among these is Amazon, the company that is most closely associated with e-commerce. Founded by Jeff Bezos, a computer science and engineering graduate from Princeton University, the Seattle-based company went online in 1995 and went public in 1997. Starting as a bookseller, the company has grown rapidly as a seller of a diverse assortment of products ranging from hardware, electronics, and music CDs to videos, toys, and tools. Revenues rose from about $150 million in 1997 to $6.9 billion in 2004. However, with the rise in revenue came a commensurate increase in operating losses, resulting in large deficits. In 2003—its ninth year of operations, and seventh after going public—Amazon finally turned a profit. Long a model for Internet companies that put market share ahead of profits and made acquisitions funded by meteoric market capitalization, Amazon.com is now focusing on making profit.

Though most of the companies in table 3.2 can be recognized by their brand names, all of them have also come to represent a particular approach in doing Internet commerce. eBay does auctions. Amazon is like an electronic mail-order company. Yahoo's search engine connects shoppers with retailers. However, because of growing competition, companies are intruding into each others' territories to add services and to attract more consumers. Amazon not only invites merchants to sell on its site and charges a commission but also has introduced a local Yellow Pages service and a Web search service that people can customize. eBay has diversified beyond the auction format by buying up e-commerce sites in India, Germany, and the Netherlands. It also bought Rent.com, an apartment-listing site, in addition to Shopping.com, the top comparison-shopping service. Yahoo originally offered links to the sites of established retailers and small merchants who paid Yahoo to host their sites; more recently, it has emphasized a search box to help shoppers find products.

Looking to the future, teleshopping will become more diverse as more and more merchants are learning to bypass eBay and Amazon by using search engine marketing to bring shoppers directly to their Web stores. One merchant explained, "eBay has been a wonderful place to start and incubate a business. Where it has been weak is helping you once you are growing and thriving. The whole Internet commerce evolution has been a fantastic opportunity that only comes along once in a generation, but now we are ready to take it to the next level."[13]

Outsourcing

The massive investment in laying underground cables that took place in the 1990s, combined with the explosion of software (such as e-mail and search engines like Google) and the spread of inexpensive computers around the world, has provided the groundwork for information to be delivered from anywhere to anywhere else, anytime. This has made possible the emergence of a completely new social, political, and business model, typically referred to as *globalization*. As discussed in chapter 2, this global restructuring reflects technological advances that have made it easier and quicker to complete international transactions beyond national borders.

Perhaps the most controversial aspect of globalization is the movement of jobs and investment from high-cost countries to lower-cost ones in order to raise profit margins. *Outsourcing* occurs when a firm subcontracts a business function to an outside supplier. Information technology has now made it possible to outsource business functions such as telemarketing, customer service, software programming, document management, medical transcription, tax preparation, and financial services, to name those that are most prominent. India has become the top destination for offshore jobs, while China is in second place but with less than half as many jobs as India.

Reputedly, it was Jack Welch, then General Electric's chairman, who started the outsourcing boom when he struck a deal with India during a sales call in September 1989. His interest was to develop a low-cost ultrasound machine for GE's medical division. The project didn't succeed as anticipated, but company executives quickly realized they had found an inexpensive pool of talented programmers and engineers. GE set aside $5 million annually for Wipro, an Indian software company, to write software for its ultrasound machines. By the mid-1990s, with orders from Welch, top GE executives began encouraging other units to follow the medical division's lead. One executive explained, "For the same amount of engineering dollars, we were getting 50 percent more people thinking about stuff worldwide."[14] Indeed a software programmer in India with two to four years' experience makes about $10,000 a year, compared with $62,000 in the United States, according to Hewitt Associates, a consulting firm.[15] By the late 1990s, GE was not only buying software from India but also had begun using the country as a base for data entry, credit card application processing, and other clerical tasks. At about this time, other companies, such as American Express and British Airways, began moving back-office operations to India (the concept of back-office operations is explained in the next section). But Welch insisted that GE go much further, shifting thousands of jobs and millions of dollars in operational expenses to India, leading to huge savings.

Critics contend that the movement of service-sector jobs abroad has contributed to the decline of U.S. manufacturing and the loss of service-sector jobs once considered safe. Available data would seem to lend support to such contention. The McKinsey Global Institute estimates that the volume of offshore outsourcing is increasing at a rate of 30 to 40 percent a year.[16] Forrester Research estimates that 3.3 million white-collar jobs will move overseas by 2015. Those sectors that are likely to be hardest hit are financial services and information technology.[17]

In contrast to those expressing alarm, others argue that passing laws to stop the export of jobs would force a higher cost base on U.S. firms, which would then lose out to foreign firms that are not so restricted—and the final cost could be less growth and the loss of even more jobs. Furthermore, they contend that an open market provides real benefits to consumers, and China will be competing aggressively with the United States as venture capitalists will go anywhere where there is low-cost labor and low-cost competition. As one observer noted, "Everything you can send down a wire is up for grabs" (see box 3.3).

Box 3.3.   More Medical and Drug Companies Do Outsourcing

At one time, it was believed that the medical and drug industries in the United States, with its heavy emphasis on scientific innovation, would be less likely to export jobs abroad than has been the case for other industries such as manufacturing and computer programming. But this is now changing. China and India are beginning to invest heavily in developing expertise in biotechnology. Singapore has been able to attract top scientists, many of them post-docs of Asian background from the United States, to establish new medical research centers. A potential advantage for these countries is in doing embryonic stem cell research, a new field that is restricted in the United States. Having completed a tour of China, Singapore, and South Korea, a group of British stem cell specialists said scientists in those countries were as competent as those in Britain, although those in the Asian countries had better equipment and funding. As they reported, "The challenge to Western preeminence in stem cell science from China, Singapore, and South Korea is real."

Other life sciences in these countries are developing as well. A recent report by Cutting Edge Information, a pharmaceutical consulting firm based in Durham, N.C., found that in the development of new drugs, American and European drug companies could cut the cost of clinical testing by more than 60 percent by going to developing countries such as India. And where Western patients are usually wary of being subject to tests, India has a huge patient population that not only offers vast diversity but is typically eager to participate.

Source: "Medical Companies Joining Offshore Trend, Too," *New York Times*, 24 February 2005.

THE TECHNOLOGICAL RESHAPING OF WORK

As society changes from an industrial to a postindustrial economy, many American workers are witnessing massive changes in their jobs, from downsizing to much greater computerization. As of 2005, only about 10 percent of the U.S. labor force was employed in manufacturing, down from 31 percent in 1960. For those workers who have failed to adjust to changing conditions, a heavy price has been paid in layoffs and other dislocations. In some instances, technology meets open resistance (see box 3.4). Most of those who have lost their jobs now require new schooling and retooling if they are to find their way back into the job market.[18]

What's more, rapid change in the economy is making traditional concepts of work obsolete, obscuring the traditional classifications of agriculture, manufacturing, and services. To a greater extent then ever before, work in all three sectors has come to consist of what can be called "mind work." Farmers now use computers to tally grain production; the UPS driver operates a computer when making deliveries; and auto mechanics use computers to make repairs. Jobs like these and many others in such fields as medicine, financial management, real estate, legal services, and advertising entail information-processing at an expanding rate.

Box 3.4.   A Gathering of Modern-Day Luddites

Some 200 years ago, a group in England called the Luddites rebelled against the Industrial Revolution by smashing machines and burning factories. Now their spiritual descendants are back. In the spring of 1996 in Barnesville, Ohio, about 350 self-proclaimed Luddites met in an old Quaker meeting house to proclaim that the so-called Information Revolution is leading to isolation and alienation where people no longer meet face-to-face.

One spokesperson, a Quaker who relied for transportation on a horse and buggy, claimed, "We're warehousing our elderly. And some people are sitting on the freeway, day after day, in a 190-horsepower steel coffin, wondering: 'Am I alone? Or does anybody else think this is crazy?'"

While insisting that he was not a Luddite, another attendee objected to what he called "Internet hucksters" who contend that being computer literate is necessary for the education of children. He rejects the claim that people without computer skills will be unemployable. Jobs, as they always have, will go to people who can get along with others, he said. "Now, how do you avoid developing those social skills? By standing at a keyboard and staring off into cyberspace for hours?"

Source: Kirk Johnson, "A Celebration of the Urge to Unplug," *New York Times*, 15 April 1996.

The physical location of work is also changing. The number of at-home workers in the United States has risen steadily since the 1980s. It is estimated that roughly 28 percent of the U.S. workforce telecommuted sometime during the average week in 2004.[19] This amounted to forty million at-home workers who were able to avoid the daily commute to their office by communicating via fax machines, cell phones, and e-mail. It is anticipated that, as gasoline prices and energy costs rise in future years, the dollar costs of operating electronic equipment at home and at alternative work sites will continue to decline, thereby increasing the relative advantage of moving various forms of production out of large central workplaces typical of the industrial era.

Maryam Alavi describes the coming of age of the "virtual" office.[20] Aided by information and wireless technologies, this relatively new way of working permits employees to work wherever it is most opportune—at home, at customers' locations, or while traveling (see box 3.5). Workers on the move—such as sales and service personnel or executives—derive special benefits from the virtual office. Field service personnel, for example, can serve their clientele better by having instant access to inventory levels, price quotes, and delivery schedules. It facilitates the locating and dispatching of service workers to customer sites. Another potential benefit is reduction of office real estate costs.

Box 3.5.   Can Vacation and Work Go Together?

A survey conducted by Travelocity, an online travel agency, found that when Americans go on vacation, many of them are inclined to take along laptops, cell phones, and video games in addition to paperbacks and bathing suits. The poll, conducted April 21–25, 2005, drew responses from 834 Travelocity members who had traveled in the previous 12 months. It was found that 88 percent of the respondents check e-mail or voice mail while traveling, and more than a third feel stressed if they can't do so. More than half of travelers between the ages of 22 and 45 admit to sneaking away from those they're vacationing with in order to check and respond to e-mail, voice mail, or instant messages.

Frank Lambert, senior technology adviser for a California software company, who travels twice a month for work and once a year for pleasure, stays connected either way. "I want to feel like my involvement in what I'm doing is important enough that it needs my participation to continue." Rob Barsky, chief operating officer of a financial investment firm agreed. "A carefree vacation is the holy grail, but it's also a leap of faith," he said. "You could come back and your job description might have changed."

Sources: "Travelocity Survey Results: Tech-Heads on Vacation," Call Center CRM News Blog, 23 June 2005, http://blog.tmnenet.com/travelocity-tech-survey-results.asp (12 July 2005); Austin Considine, "Away from It All? Not Quite," *New York Times*, 3 July 2005.

In addition to working from home or an alternative workplace such as a neighborhood telework center, some people can work at a distance but not necessarily outside of an office. Generally referred to as "distributed work," it includes group activities such as video conferencing and networked information resources that make it possible for people from far-away locations to work together. Distributed work can cut travel costs and permit work to be done that would otherwise require great expense or inconvenience.[21]

Alvin Toffler paints a cozy picture of electronic cottages that will restore the family as a functioning group. As he describes it, "The electronic cottage raises once more on a mass scale the possibility of husbands and wives, and perhaps even children, working together as a unit."[22] As more people work at home, furthermore, it will allow them to change jobs without changing houses. With less forced mobility, according to Toffler, we can expect greater community stability and an increase in participation in voluntary organizations such as churches, lodges, clubs, and youth organizations.

Other observers are more critical. They contend that working at home encourages long and irregular hours and removes the possibility of getting a respite from the workplace. In addition, employers are beginning to show concern over the generally unstructured and often inefficient work habits of telecommuting employees. Box 3.6 illustrates how one firm, Merrill Lynch, is attempting to regulate working at home by suggesting rules for how to work, when to work, and how to work safely.

More imminent are the effects of new technologies on business practices. A key consideration, now and in the future, is the degree to which different business operations require spatial proximity to suppliers, customers, competitors, and other units in the firm. Where work involves manipulation of physical objects (such as machines in auto repair, food in restaurants, hair in beauty shops), locations would continue to be bound by the need for contact with customers. Also, some work operations may be automated, as in self-service check-in and check-out hotels. In such cases, services would stay close to the customer, though it is likely that jobs would be cut.

However, where work involves an exchange of physical items with the computer (e.g., wholesaling or check processing), the informational aspects of these functions could be carried out over considerable distance via telecommunications. In the past, most firms inscribed information on paper and transferred communiqués physically, necessitating the employment of file clerks, typists, and messengers. Today, electronic imaging permits data to be transmitted electronically instead of on paper, and consequently more and more firms are using computer-based

Box 3.6.   How to Be a Telecommuter

	Old Way	*Merrill Lynch Way*
How to begin	Find a laptop from the office storage closet. Take it home. Work.	Invest hours in formal instruction and at least two weeks in simulation lab.
Workstation	Place lounger on back patio. Tape TV section in place to reduce screen glare.	Solicit company approval of chair, desk, and lighting as ergonomically correct.
Child care	Look to sneak in some work during nap time and *Sesame Street*.	Show proof of child care during work hours.
Workday	Respond to questions about workday whereabouts with vague answers about having just stepped out for a minute.	Submit schedule in writing for approval of two to four workdays at home per week.
Communicating	Wrestle with a busy signal for hours to ask the technical department why home computer no longer works properly.	Call the twenty-four hour-a-day help desk during off-hours for advice and technical support.
Follow-up	Feel humiliated when the boss doesn't recognize you at the office Christmas party.	Conduct regular surveys, focus-group meetings, and mandatory office appearances.

Source: Kirk Johnson, "Limits on the Work-at-Home Life," *New York Times*, 17 December 1997.

systems for almost all forms of communication. As a general trend, telecommunication has greatly increased the range of services that can be conducted without requiring the physical presence of customers. For example, most insurance companies file customer information in a centralized location, which allows one customer-service representative to view the customer's entire history of interaction with the company. Credit cards and toll-free telephone services also allow customer-service functions to be conducted over the phone from centralized service centers. Thus, more and more business transactions are shifting from face-to-face communication and interaction to cheaper electronic forms of communication (see box 3.7).

Box 3.7.   Types of Linkages between Work Operations

Linkage	*Means*
Face-to-face	In-person meetings and interaction
Voice and video	Video conferencing and video phones
Voice	Telephone, voice mail, cell phone
Electronic data	E-mail, fax, electronic data technology
Physical mail	Postal service, overnight mail
Goods shipment	Conventional freight movement

INFORMATION TECHNOLOGY AND
THE EMERGENCE OF NEW SPATIAL PATTERNS

As we have already noted, American cities evolved as centers of business and commerce out of the need for physical proximity among firms, suppliers, and customers. Agglomerations of people and industry made for efficiency in the production and distribution of goods and services. During the industrial era, new transportation modes (e.g., trains, electric trolleys, cars, and trucks) allowed some separating out in shaping the mass-production city. In forming the postindustrial city, information technology makes possible the large-scale separation of a broad range of functions.

It should be emphasized, however, that the spatial reordering of business and commercial activities depends largely on the type of functions being performed. So-called front-office functions, requiring direct contact with customers, are driven by considerations of location and are different from routine back-office functions that involve no direct customer contact. Special considerations also apply to activities involving goods and production distribution as well as to complex office work.

Front-Office Operations

As a general rule, the location of a major share of service jobs is determined by local market demand. This includes such services as branch banks, retail stores, beauty shops, and auto repair shops, among others. These are usually located where customers can have direct access. However, developments in information technology, and particularly in data-based access, have increased the share of services that can be centralized without consideration of physical proximity to the customer. For example, banks have moved loan and credit card processing out of local branches to centralized customer-service centers. Telephone technologies have made it possible for telemarketing operations to be located in places such as Omaha and San Antonio, which have become centers for a large number of telemarketing firms. In addition, major banks, software companies, and information service companies can be accessed by customers directly from home. By removing service operations from local sites, firms increase economies of scale in production. The transition away from local front-office functions is likely to continue as consumers of services grow more comfortable handling a wide range of transactions without face-to-face contact.[23]

Routine Back-Office Functions

Most back-office work involves routine information processing apart from direct dealing with customers. Until recently, back-office activities usually occurred behind the front office and were located in office buildings in the central business district of older central cities. There were many reasons for this. As the transportation hub of the region, the city was the best location for bringing together large numbers of workers. In addition, the large volume of paper transactions necessitated proximity with both front-office and back-office managerial functions. "The back office was like an assembly line where paper was processed and information added at certain places (the way parts are added to a car during assembly)."[24] As the volume of transactions increased over the years, buildings in the central business district rose higher and higher to provide needed space, and cities such as New York and Chicago became the skyscraper capitals of the world.

E-mail, teleconferencing, fax machines, and wireless communications have made it possible for companies to spin off routine functions from complex functions and, using cost-benefit criteria, to place each in its optimal location. Consequently, the central business districts of cities are being transformed as many back-office operations are relocated to more distant areas where costs are lower. Functions that utilize lower skills and a high percentage of clerical workers, such as telemarketing, are more likely to be located in smaller towns and cities.

As part of a growing trend, furthermore, other back-office functions are being located overseas. Several insurance companies in the United States have established processing operations in Ireland. Other companies have set up call centers and help desks in such countries as India, China, and Mexico. These companies take advantage of relatively low wage rates for workers with good computer skills and a lower employee turnover rate. To illustrate, it is estimated that approximately 245,000 Indians are presently employed answering phones or dialing out to solicit people for credit cards or nonpayment of bills. In many instances, they are given

---

Box 3.8.   What an Indian Call Center Looks Like

Do you know what an Indian call center looks like?

While filming [a] documentary about outsourcing, the TV crew and I [Thomas Friedman] spent an evening at the Indian-owned "24/7 Customer" call center in Bangalore. The call center is a cross between a co-ed college frat house and a phone bank raising money for the local public TV station. There are several floors with rooms full of twenty-somethings—some twenty-five hundred in all—working the phones. Some are known as "outbound" operators, selling everything from credit cards to phone minutes. Others deal with "inbound" calls—everything from tracing lost luggage for U.S. and European airline passengers to solving computer problems for confused American consumers. The calls are transferred here by satellite and undersea fiber-optic cable. Each vast floor of a call center consists of clusters of cubicles. The young people work in little teams under the banner of the company whose phone support they are providing. So one corner might be the Dell group, another might be flying the flag of Microsoft. Their working conditions look like those at your average insurance company. Although I am sure that there are call centers that are operated like sweatshops, 24/7 is not one of them.

Source: Reprinted by permission of the publisher from Thomas L. Friedman, *The World Is Flat: A Brief History of the Twenty-first Century* (New York: Farrar, Straus and Giroux, 2005), 21, 22.

American names and trained to speak with an American accent. When on the phone, Americans think that they speaking to someone in the United States rather than to someone in another part of the world (see box 3.8).

Not too long ago, customer service representatives in most telephone companies had to access paper records to respond to inquiries; now, any service representative in the company can gain online access to any customer's record in any location. Consequently, under pressure to reduce costs, a pattern of centralization and decentralization has resulted. AT&T has reorganized itself into six megacenters for its residential customer-service line, plus several more for small businesses, network management, and call-completion operations. Companies are centralizing into megacenters in order to gain economies of scale—cutting back on employees as well as on other costs. Relatively new competitors such as Sprint and MCI built megacenters into their networks from the beginning, locating them in lower-cost regions of the country. Sprint, for example, has centers in Jacksonville, Dallas, Kansas City, Phoenix, and Winona, a small city in Minnesota. Other prominent examples include the transfer of back offices at American Express from New York to Salt Lake City, Fort Lauderdale, and Phoenix, and the transfer of back-office operations at Metropolitan Life to Greenville, Scranton, and Wichita.

Goods Production and Distribution

There are four main components in the processing of goods: production, transportation, distribution, and sales. In terms of production, it is necessary to distinguish between routine production and complex production. As part of a growing trend, manufacturing firms are relocating routine low-skill assembly and warehouse functions to low-cost areas where telecommunications allow interaction between distant headquarters and branch facilities.[25] However, where the production process is less routine and requires more intricate coordination, centralized locations closer to markets and suppliers are still favored.

In chapter 2, we noted the decline of manufacturing in older regions and cities in the United States. This trend began in New England in the mid-1970s and spread to the rest of the Northeast and Midwest in the late 1970s and early 1980s. Many manufacturing firms moved out of Frost Belt cities to relocate in the newer environs of the Sun Belt, and in many cases, as previously noted, manufacturing left the United States entirely for low-wage developing countries. Whereas manufacturing had initially located in cities with access to energy, ports, railroads, and markets, today's advancements in shipping technology, commercial aviation, interstate highways, and large refrigerated trucks have obviated industry's need to remain in the central city.

In contrast to the spatial features of goods production, facilities for the wholesaling and distribution (freight) of goods tend to be located outside the core of large metropolitan areas where land and labor costs are lower. This is made possible by information technology that allows firms to deliver goods faster than ever before, thereby enabling them to avoid the higher costs of locating in large urban areas. Similarly, information technology allows freight transportation functions to consolidate and serve wider markets from fewer locations.

There are some important exceptions to these patterns. For example, some manufacturing businesses—such as printing, food processing, construction materials, and arts/entertainment equipment—tend to remain in cities to serve local markets. The importance of design as an art form also helps to explain a modest revival of certain types of manufacturing in cities such as New York and Los Angeles. Textile manufacturing in these places has received a boost from upscale fashion requirements in the garment industry. In addition, the need for close collaboration with other firms, particularly service firms, may give certain urban core locations a competitive advantage. The movie and television industry in Los Angeles consists of many small firms linked regionally with corporate distributors at one end and various suppliers on the other.[26] New York's role as a center for the arts cultivates customized manufacturing for niche markets in such fields as entertainment, fashion, and design.

Complex Office Work

Though cyberspace continues to reduce the need for spatial proximity in many fields of endeavor, not all functions are amenable to its workings. Many functions still depend on close face-to-face interaction. They include types of activities that are nonroutine and complex in nature and typically involve managers, professionals, and executives in specialized work such as accounting, consulting, legal work, research and development, and corporate headquarters offices. Although information technology is utilized in complex office work, it does not substitute for close physical contact as a necessary ingredient of higher forms of professional decision making.[27]

In his research, Sukkoo Kim points out that 40 percent of U.S. employment is packed into 1.5 percent of the country's land area—and those cities that specialize in finance, insurance, property dealing, and wholesale trade tend pack more and more workers onto the same land.[28] Of course, some cities are able to accommodate face-to-face interaction and networking better than others. Places such as Wall Street in Manhattan, San Francisco's Financial District, or the Loop in Chicago are better able to facilitate the transfer of privileged information that is so critical to high-level decision making, while the downtowns of other cities such as Detroit, Cleveland, and Baltimore appear to have lost that capability. (This is discussed in greater detail in chapter 4.)

Large, densely settled cities that facilitate interpersonal networking also rank high as centers of innovation. Firms located in the urban core usually have access to specialized skills, detailed market data, and the newest technologies for the development of new products. Manuel Castells and Peter Hall have adopted the French term *technopoles* to describe science-park clusters that concentrate technologically innovative, industry-related production.[29] The most often-cited example is an area lying between San Francisco and San Jose in what has come to be known as Silicon Valley, home of the largest pool of venture capital on Earth. Other high-tech centers are the Route 128 corridor encircling Boston and the Research Triangle Park near Raleigh, North Carolina. The creation of high-tech centers in metropolitan areas is driven in part by the need for technologically based manufacturers to interact closely with suppliers, customers, universities, and research institutes. Because of rapidly changing technology and markets, moreover, cooperation among firms is becoming more and more important, and such forms of cooperation are enhanced where firms locate in clusters.

SUMMARY AND CONCLUSIONS

Beginning slowly in the 1940s and 1950s with the development of the first electronic computers and moving much more rapidly toward the end of the century with the development of highly sophisticated silicon chips, electronic communications have been replacing traditional methods of conducting transactions (see box 3.9). Because a growing share of societal activities consists of transmitting information—be it bank data, demographics, inventories, client records, or delivery schedules—the potential of information technology to shape spatial patterns in almost all aspects of the way people live, work, and play is now greater than ever before.

Industries that in the past would have clustered in the urban core near transportation hubs are now connecting over vast spatial expanses and

---

Box 3.9. A Brief History of Communications Technology

1844	Samuel Morse telegraphs a message from Washington, D.C., to Baltimore.
1858	Telegraph cable laid across the Atlantic.
1876	Alexander Graham Bell invents the telephone.
1879	Thomas Alva Edison invents the lightbulb.
1896	Guglielmo Marconi sends a radio signal.
1935	First electric typewriter sold.
1946	First television program broadcast coast-to-coast.
1947	Transistor invented at Bell Labs.
1950	Semiconductor chip invented.
1969	Defense Department begins ARPANET, forerunner of the Internet.
1973	Dr. Martin Cooper, at the systems division of Motorola, makes the first call on a portable wireless phone.
1977	Bell Labs constructs and operates a prototype cellular phone service.
1981	Laptop computer invented by Adam Osborne.
1982	The Federal Communications Commission authorizes commercial cellular service for the United States.
1982	Commercial electronic mail introduced, with service among twenty-five cities.
1989	Tim Berners-Lee proposes World Wide Web project.
1993	First World Wide Web conference held in Cambridge, Massachusetts.
1994	Jim Clark and Marc Andreessen form Mosaic Communications, forerunner of Netscape.
1999	Canadian firm RIM develops a handheld wireless e-mail messaging computer called the BlackBerry.

can move ever greater distances from the metropolitan center. Workers can telecommute to work and students can receive video instruction outside the classroom, either at home or at specified sites. Using the Internet, researchers can access information from almost anywhere at any time, day or night. Video conferencing allows professionals of all kinds to participate in discussions and consult without having to move from their home offices. And banking and finance are no longer locally oriented but, through electronic means, can exert influence throughout the world.

All this movement of activity to cyberspace is forcing a reconsideration of traditional understandings of community. "Community" may refer to a place or group of people sharing space, such as a legally constituted municipality, or it can mean a group that shares common values, culture, or interests but not necessarily a geographic location. Another type, often called a traditional community, is a group that shares a common territory as well as culture, origins, and identity. In his book *City of Bits*, William J. Mitchell contends that concepts of traditional community are changing and that a community may now find its place in cyberspace. The new urban site, he says, is not some patch of earth but a computer to which members may connect from wherever they happen to be. Looking to the future, more and more virtual communities will be created within which "social contracts will be made, economic transactions will be carried out, cultural life will unfold, surveillance will be enacted, and power will be exerted."[30]

In light of all this, we must consider what will happen to traditional cities as we have known them. What new urban forms are evolving and what will cities look like in the twenty-first century?

NOTES

1. Bill Gates, with Nathan Myhrvold and Peter Rinearson, *The Road Ahead* (New York: Viking, 1995).

2. Martin Gay and Kathlyn Gay, *The Information Superhighway* (New York: Twenty-first Century Books, 1996), 3.

3. D. N. Hatfield, *The Technological Basis for Wireless Communications* (Queenstown, Md.: Institute for Information Studies, 1996).

4. Paul Levinson, *Cellphone: The Story of the World's Most Mobile Medium and How It Has Transformed Everything* (New York: Palgrave Macmillan, 2004).

5. Katie Hafner and Matthew Lyon, *Where Wizards Stay Up Late: The Origins of the Internet* (New York: Simon & Schuster, 1996).

6. William J. Mitchell, *City of Bits: Space, Place, and the Infobahn* (Cambridge, Mass.: MIT Press, 1995).

7. See Alberto Manguel, "The Pursuit of Knowledge, From Genesis to Google," *New York Times*, 19 December 2004.

8. B. Kurshan and C. Lenk, "The Technology of Learning," in *Crossroads on the Information Highway: Convergence and Diversity in Communications Technologies* (annual review of the Institute for Information Studies, Aspen Institute, and Northern Telecom; Queenstown, Md.: Institute for Information Studies, 1995), 31–50.

9. J. F. Moore, "Convergence and the Development of Business Ecosystems," in *Crossroads on the Information Highway: Convergence and Diversity in Communications Technologies* (annual review of the Institute for Information Studies, Aspen Institute, and Northern Telecom; Queenstown, Md.: Institute for Information Studies, 1995), 51–61.

10. R. Tompkins, "Shop-Till-You-Drop at the Touch of a Button," *Financial Times*, 9 June 1994.

11. Judy Hillman, *Telelifestyles and the Flexicity: A European Study—The Impact of the Electronic Home* (Dublin: European Foundation for the Improvement of Living and Working Conditions, 1993), 2.

12. Leslie Walker, "E-Commerce's Growing Pains," *Washington Post*, 25 June 2005.

13. Walker, "E-Commerce's Growing Pains."

14. Jay Solomon and Kathryn Kranhold, "In India's Outsourcing Boom, GE Played a Starring Role," *Wall Street Journal*, 23 March 2005.

15. Solomon and Kranhold, "In India's Outsourcing Boom."

16. This estimate is cited in Daniel Drezner, "The Outsourcing Bogeyman," *Foreign Affairs* 83, no. 3 (May–June 2004): 24.

17. Drezner, "The Outsourcing Bogeyman," 24.

18. E. E. Vogt, "The Nature of Work in 2012," in *Crossroads on the Information Highway: Convergence and Diversity in Communications Technologies* (annual review of the Institute for Information Studies, Aspen Institute, and Northern Telecom; Queenstown, Md.: Institute for Information Studies, 1995), 80–91.

19. Cahners In-State Group, "Number of Telecommuters Swells to 32 Million in 2001," In-Stat.com, 2002, http://goliath.ecnext.com> (October 3, 2005).

20. Maryam Alavi, "Dick Tracy's Office—Business Applications of Wireless Technologies," in *The Emerging World of Wireless Communications*, ed. C. M. Firestone (Queenstown, Md.: Aspen Institute and Institute for Information Studies, 1996).

21. J. H. Pratt, "Home Teleworking: A Study of Its Pioneers," *Technological Forecasting and Social Change* 25 (1984): 1–14.

22. Toffler, *The Third Wave* (New York: Morrow, 1980), 190–98; quote on 190.

23. U.S. Congress, Office of Technology Assessment, *The Technological Reshaping of Metropolitan America* (Washington, D.C.: GPO, 1995), 109, 110.

24. U.S. Congress, *Technological Reshaping*, 110.

25. A. J. Scott, "Flexible Production Systems: The Rise of New Industrial Spaces in North America and Western Europe," *International Journal of Urban Regional Research* 2 (1988): 171–85.

26. S. Christopherson and M. Storper, "The Effects of Flexible Specialization on Industrial Politics and the Labor Market: The Motion Picture Industry," *Industrial and Labor Relations Review* (1989): 331–47.

27. See Donald Trump, *The Art of the Deal* (New York: Time Warner, 1987).

28. Cited in "Press the Flesh, Not the Keyboard," *Economist* (August 22, 2002): 50.

29. Manuel Castells and Peter Hall, *Technopoles of the World: The Making of Twenty-First-Century Industrial Complexes* (New York: Routledge, 1994).

30. Mitchell, *City of Bits*, 160.

4

Types of Cities

As noted in the previous chapters, new information technologies are dramatically changing the economies of American cities. By allowing people to communicate (through telework, telecommerce, tele-information, teleculture) with no regard for distance (whether within metropolitan areas, between regions, or between continents), business organizations thousands of miles apart can shut down old industries or open new ones; they can shift thousands of workers from one location to another; and, with the press of a button, they can instantaneously transfer huge investments to any-where in the world. Taking advantage of their newfound mobility, many firms and even whole industries now have the capacity to locate wherever they want. This creates fierce competition among cities to attract jobs and investment; and while some cities succeed and become winners, others struggle and lose, and still others accept their status as mere survivors.

As referred to in chapters 2 and 3, an important element in accounting for the transformation of cities is globalization. Writing in the early 1990s, Robert Reich weighed the implications of this process:

> We are living through a transformation that will rearrange the politics and economics of the coming century. There will be no national products or tech-nologies, no national corporations, no national industries. There will no longer be national economies, at least as we have come to understand that concept. . . . As almost every factor of production—money, technology, fac-tories, and equipment—moves effortlessly across borders, the very idea of an American economy is becoming meaningless, as are the notions of an Amer-ican corporation, American capital, American products, and American tech-nology.[1]

Of special significance in the globalization process has been the emergence of the transnational corporation. Fiber-optic networks and satellite communications systems have enabled firms to handle enormous volumes of financial and business transactions around the globe. Paul I. Knox refers to "commodity chains" that crisscross global space. As he explains it:

> Commodity chains are networks of labor and production processes whose origin is in the extraction or production of raw materials and whose end result is the delivery and consumption of a finished commodity. They are, effectively, global assembly lines that are geared to produce global products for global markets. These assembly lines often span countries and continents, linking the production and supply of raw materials, the processing of raw materials, the production of components, the assembly of finished products, and the distribution of finished products into vast webs of interdependence.[2]

According to Knox, the phenomenon of constantly changing chains of economic interdependence serves as context for understanding the changing roles of cities. Some cities, having strategically positioned themselves by effectively linking whole complexes of commodity chains, assume the role of global cities. Critical factors in helping them to attain such status is their ability to provide the necessary support services such as transportation and communication networks, a workforce with the requisite technical and professional skills, and strong research and development institutions. Most important, such cities have succeeded in establishing themselves as high-powered command centers in the organization of the world economy (see box 4.1).

Most American cities have come to perform more marginal roles in the global economy. Some are export centers for low-value-added, labor-intensive products made by unskilled workers. Others have lost their manufacturing base and now strive to restructure their economies by aggressively recruiting high-tech industries. Given the increased mobility of capital and the limited capability of government to enforce development conformity among localities, urbanization now reflects a growing diversity of functional capacity and development. According to John Logan and Harvey Molotch, because some urban areas are better able than others to attract investment, metropolitan economies reflect different growth patterns.[3] In light of this, they view urban places as falling into one or more of the following categories:

1. Headquarters cities
2. Innovation centers
3. Old industrial cities
4. Border cities
5. Retirement sites

Box 4.1. Cities in a World Economy

A central thesis [by the author] . . . is that the transformation during the last two decades in the composition of the world economy accompanying the shift to services and finance brings about a renewed importance of major cities as sites for certain types of activities and functions. In the current phase of the world economy, it is precisely the combination of the global dispersal of economic activities *and* global integration—under conditions of continued concentration of economic ownership and control—that has contributed to a strategic role for certain major cities that I call *global cities*. Some have been centers for world trade and banking for centuries, but beyond these long-standing functions, today's global cities are (1) command points in the organization of the world economy; (2) key locations and marketplaces for the leading industries of the current period, which are finance and specialized services for firms; and (3) major sites of production for these industries, including the production of innovations. Several cities also fulfill equivalent functions on the smaller geographic scales of both trans- and subnational regions.

Alongside these new global and regional hierarchies of cities is a vast territory that has become increasingly peripheral, increasingly excluded from the major economic processes that fuel economic growth in the new global economy. A multiplicity of formerly important manufacturing centers and port cities have lost functions and are in decline, not only in the less developed countries but also in the most advanced economies.

Source: Reprinted by permission of the publisher from Saskia Sassen, *Cities in a World Economy* (Thousand Oaks, Calif.: Pine Forge Press, 1994), 4.

Two more types could be added to the list:

6. Leisure-tourism playgrounds
7. Edge cities

These types are, at best, approximate descriptions of real cities, and in many instances, they overlap. Nor do they represent an exhaustive typology. Rather, they are intended as general categories for suggesting new forms of urban development with particular consequences for residents in terms of wages and wealth, taxes and services, and quality of life.

HEADQUARTERS CITIES

Headquarters cities may also be referred to as "world" cities, "global" cities, or sometimes "capital" cities. They are places where banks, corporate headquarters, and other command functions and high-level producer service entities, such as law firms and advertising agencies, are concentrated. At the highest level, they manifest certain characteristics:

- They are sites of most of the leading global markets for commodities and investment capital, foreign exchange, equities, and bonds.
- They are the sites with the highest concentration of corporate headquarters, including transnational corporations and large foreign firms.
- They are the locations of national and international headquarters of trade and professional associations.
- They are the locations of choice for national and international media organizations and news and information services.
- They are usually major cultural capitals for the arts and design, fashion, film, and television.

Saskia Sassen contends that global cities have emerged as strategic sites in the world economy. Decisions made in New York, London, or Tokyo affect jobs, wages, and the economic health of locations throughout the world.[4]

As noted in chapter 2, in the United States, New York stands at the peak of the urban global system, followed by such second-tier cities as Chicago and Los Angeles. But Peter J. Taylor and Robert E. Lang ask: What about other cities? How do they place in the global system of economic relations?[5] Using network analysis, their basic thesis is that if cities are to grow, they must attract globally connected, high-value service firms. In light of this, they track the global distribution of one hundred leading advanced service firms to determine cities' economic connectedness to other world cities and the patterns of these linkages across the globe. They explain their methodology as follows:

> Cities are not themselves the prime agents of world city network formation, however; it is advanced producer service firms that have been largely responsible for creating and maintaining the network. These firms have offices in important cities across all world regions, and personnel, information, knowledge, intelligence, ideas, plans, instructions, and advice flow freely among them. As such, these global service firms "interlock" the cities in which they have a presence. Viewed this way, the world city network can be measured.[6]

Overall, Taylor and Lang identified 100 such firms—18 in accounting, 15 in advertising, 23 in banking/finance, 11 in insurance, 16 in law, and 17 in management consulting—across 315 cities worldwide. They then measured "products of service values" for pairs of cities as indicating connection potential. Summing all such products for a particular city for all firms across all other cities serves as the criterion for defining the city's *global network connectivity* (GNC).

Table 4.1 shows the GNC scores for U.S. cities, presented by size of the GNC and grouped into ten strata based on their connectivity levels. As expected, New York in Stratum I and Chicago and Los Angeles in Stratum II are ranked at the top of the world city network.[7] Because of its key role with respect to trade and finance with Latin America, Miami is in the third tier. Other cities in the third stratum are San Francisco, which serves as the financial center of the West; Atlanta, which functions as the unchallenged media and financial capital of the rapidly growing South; and Washington, the nation's capital. Lower down in the fourth stratum are Boston and Seattle, which serve primarily as regional centers in New England and the Pacific Northwest, respectively, and two Texas cities—Dallas, the financial center of the Southwest, and Houston, the world's energy capital. At the bottom of the global system, in strata IX and X, are Las Vegas, New Orleans, Sacramento, Omaha, and Wilmington, Delaware. As we shall see next, such cities survive by performing other roles in the U.S. economy. One finding that was not expected is that Phoenix and San Jose, two cities that have shown dynamic growth over the past score years, have relatively little connectivity. To convey additional insight into the phenomenon of global cities, we will take a more in-depth look at two prime examples: the New York and Los Angeles regions.

New York City and Environs

An overview of New York City's economic status during the past thirty years reveals an interesting paradox. While the city qualifies as a world-class financial capital along with Tokyo and London, its local economy appears vulnerable. Close to bankruptcy in the 1970s, New York has been subject to heavy employment losses in manufacturing and to shrinkage in wholesale and retail trade (see figures 2.2, 2.3, and 2.4). The national recessions of 1969, 1974, and 1990 were especially damaging, and the city was shaken to the core during the stock market collapse in 1988. To be noted also, as shown in table 4.2, is that there are now more Top 500 corporate headquarters in the surrounding suburbs than in New York City proper. Pessimists have no difficulty finding the dark lining in the city's economy.

Nevertheless, beginning in the 1970s, New York was poised to take advantage of new development forces that began to emerge in the international economy. In particular, increases in the mobility of capital, particularly across international borders, came to be of growing importance. By virtue of its dual international and national orientation, New York has been able to reap the benefits of these new global forces. In addition,

Table 4.1. Global Network Connectivities of U.S. Metropolitan Areas

City	GNC Score*	Stratum
New York	61,895	I
Chicago	39,025	II
Los Angeles	38,009	II
San Francisco	32,178	III
Miami	29,341	III
Atlanta	27,052	III
Washington, D.C.	26,522	III
Boston	22,249	IV
Dallas	21,796	IV
Houston	21,424	IV
Seattle	19,252	IV
Denver	17,368	V
Philadelphia	17,006	V
Minneapolis	16,914	V
St. Louis	16,124	V
Detroit	15,818	V
San Diego	14,585	VI
Portland, Ore.	14,113	VI
Charlotte	13,556	VI
Cleveland	13,442	VI
Indianapolis	13,347	VI
Kansas City	12,772	VI
Pittsburgh	12,707	VI
Baltimore	11,309	VII
Phoenix	11,025	VII
Cincinnati	10,603	VII
Tampa	10,532	VII
Columbus	9,974	VII
San Jose	9,843	VII
Rochester	9,731	VII
Palo Alto	9,078	VIII
Hartford	9,007	VIII
Richmond, Va.	8,845	VIII
Buffalo	8,798	VIII
Honolulu	8,656	VIII
Las Vegas	7,911	IX
New Orleans	7,089	IX
Sacramento	6,870	IX
Omaha	6,564	IX
Wilmington, Del.	3,740	X
Other cities	—	X

*GNC = global network connectivity
Source: Peter J. Taylor and Robert E. Lang, "U.S. Cities in the 'World City Network'" (Washington, D.C.: Brookings Institution, 2005), available at www.brookings.edu/metro/pubs/20050222_worldcities.pdf.

Table 4.2. Headquarters Location of 500 Largest Publicly Held Corporations in the United States

Metropolitan Area	Number of Corporations	Number in Cities/ Surrounding Suburban Areas
New York	76	29/47
Chicago	35	17/18
San Francisco	29	4/25
Los Angeles	25	5/20
Boston	19	2/17
Dallas	19	13/6
Houston	18	18/0
Minneapolis	14	11/3
Washington, D.C.	12	4/8
Philadelphia	11	3/8

Source: Mark Abrahamson, *Global Cities* (New York: Oxford University Press, 2004), 17. Abrahamson derived data from *Ward's Business Directory*, vol. 4 (Detroit: Gale, 1999). Figures pertain to 1998.

international trade and global financial investment activities have served to promote such other areas as marketing and advertising, finance and banking, broadcasting, information technology, publishing, and real estate.

Thus, it can be argued that a sufficient critical mass of activities remains in the city and the region. When the banks and security firms on Wall Street do well, the benefits diffuse into accounting, legal services, insurance, advertising, and publications—and go indirectly to restaurants, theaters, and entertainment that serve the high life. While advances in telecommunications have facilitated moving back-office activities out of New York, at the same time the city has become the site of a new high-tech sector—Silicon Alley, a conglomeration of companies developing multimedia software, Web sites, online entertainment, and related goods and services.

New York, moreover, continues to exude an aura of cultural vitality that attracts the best talents from around the world. The city's cultural industry has profited from its association with corporate management and through the patronage of wealthy benefactors.[8] New York is also one of the nation's most popular destination for tourists, which reinforces the city's preeminence in theater, music, dance, and the visual arts.

While New York's attraction for immigrants is generally viewed as a mixed blessing, burdening the school system and social services, it can also be argued that this constant renewal of the city's energy outweighs the costs. By moving into neighborhoods that were being vacated because of deindustrialization, immigrants have helped to stabilize the city's communal infrastructure. The city's foreign-born population, which was just 18 percent in 1970, reached 38 percent in 2000 and is still rising rapidly with Dominicans, Guyanans, Russians, Chinese, Jamaicans, and Indians leading the charge. These immigrants, furthermore, contribute to the city's role as an international financial center. Their ability to understand and speak foreign languages makes it easier to accommodate foreign nationals.

Though New York continues to lead as an international city, its internal economy manifests certain structural weaknesses that are a continuing source of concern to its leadership. One concern is the precarious nature of the city's economy, which is based largely on the securities industry. When Wall Street prospers, the city prospers. This was true in the late 1990s when average wage income increased 7.6 percent, half-vacant office buildings in Manhattan were filling up, and the city budget was balanced for the first time in many years. When the securities industry falls, however, as it did in October 1987 and again in May 2001, all of this goes into reverse, raising the threat of fiscal deficits at City Hall. Even in prosperity, the city must contend with a huge bill for deferred maintenance: tens of billions of dollars are needed to bring schools, roads, bridges, and other infrastructure up to standards.

Another concern refers to the totally unexpected tragic events that took place on September 11, 2001, when two hijacked commercial jetliners were crashed into the upper stories of the Twin Towers of the World Trade Center, leading to their collapse. Many surrounding buildings were severely damaged, and more than 2,700 people lost their lives. Of the 2,198 nonrescue workers that were killed, 78 percent were employed in finance, insurance, and real estate. The effects of the attack, along with a weakening national and global economy, helped to create an extremely volatile economic environment. Within the city, the attack resulted in approximately 430,000 lost job months and a loss in wages of $2.8 billion. As reported by the U.S. Bureau of Labor Statistics, the effect of 9/11 was centered on the city's "export" sector—the most internationally oriented part of the economy.[9]

Still another concern is that, like many other older central cities, New York suffers from a wide gap between rich and poor. This is explained largely by the strong demand for skilled workers and weak demand for the unskilled as well as an oversupply of people competing for those jobs. The many manufacturing jobs that have been lost in the city have largely affected unskilled workers. Roughly two-thirds of all growth in wages during recoveries from recent economic downturns has been in financial services, and all the benefits go to skilled workers and top management. For the bottom 20 percent of the city's households, income levels have remained static.

John Mollenkopf and Manuel Castells refer to the income inequality in New York as creating a "dual city" containing two opposing forces: The

---

Box 4.2.   The "Dual City" Metaphor

New York incontestably remains a capital for capital, resplendent with luxury consumption and high society, as *Town and Country* proclaimed in a cover story on the "empire city." But New York also symbolizes urban decay, the scourge of crack, AIDS, and homelessness, and the rise of a new underclass. Wall Street may make New York one of the nerve centers of the global capitalist system, but this dominant position has a dark side in the ghettos and barrios where a growing population of poor people lives.

These trends have prompted the literary imagination to embrace the "dual city" metaphor, as in Tom Wolfe's best-selling *Bonfire of the Vanities.* In it, Sherman McCoy, a Wall Street "master of the universe," is brought low by his contact with the mean streets and political byways of the South Bronx. Nor is the appeal of the "dual city" metaphor merely journalistic or literary. *New York Ascendant* (1988), the report of the city's Commission on the Year 2000, expressed a similar view. Noting that "New York's poverty is not new," it found that "today's poor live in neighborhoods segregated by class with few connections to jobs . . . a city that was accustomed to viewing poverty as a phase in assimilation to the larger society now sees a seemingly rigid cycle of poverty and a permanent underclass divorced from the rest of society."

Source: John Hull Mollenkopf and Manuel Castells, "Introduction," in *Dual City: Restructuring New York*, ed. John Hull Mollenkopf and Manuel Castells (New York: Russell Sage Foundation, 1991). Reprinted with permission.

---

professionals of the corporate sector constitute a social network whose interests are driven by the development of New York City's corporate economy, while the remaining social strata represent diverse interests and values (see box 4.2). "Neighborhood life thus becomes increasingly diverse and fragmented by race, ethnicity, gender, occupational and industrial location hindering alliances among these groups."[10]

Greater Los Angeles

Los Angeles is the second largest city in the United States and the economic financial hub of the western United States (see box 4.3). Comparing Los Angeles with New York, major differences are revealed. For one thing, New York has many more headquarters of Fortune 500 companies and consequently has much more office space in its central core. Because Los Angeles charges a gross receipts tax based on a percentage of business revenue, companies benefit by locating in neighboring cities that charge only small flat fees. In addition, as a relatively young city, Los Angeles has experienced rapid development in the twentieth century. Growing from 1.3 million people in 1940 to 3.7 million in 2000—9.8 million in Los Angeles

---

Box 4.3. Overview of Greater Los Angeles

Greater Los Angeles is the suburban hinterland of the City of Los Angeles. It is such a sprawling area that residents typically identify with broad subregions. Though outsiders commonly refer to the entire region as "L.A.," it contains five counties, more than one hundred municipalities, and hundreds of neighborhoods and districts. As of 2005, the official estimate of the population of Greater Los Angeles is 17,545,623.

Los Angeles first developed as a city when automobiles began to be mass-produced, which contributed to urban spread. This has resulted in the City of Los Angeles having very low population density compared to other large cities, though the metropolitan area as a whole has a relatively high density of 7,070 people per square mile. The metropolitan area's vast freeway system has made it the archetypal auto-dependent urban area. The huge number of motor vehicles, combined with the city's valley location, often creates dangerously high smog levels.

The City of Los Angeles has one of the busiest ports in the United States, with roughly half of its commerce coming from other nations, and its international airport is one of the world's busiest. In Los Angeles alone, more than 2,000 cars are sold daily, including 20 percent of all U.S. Rolls-Royce registrations. At the same time, however, the city is known as the "homelessness capital" of the United States, and one in seven of the county's residents relies on some form of public assistance.

County—the city and its region have come to represent unbridled urban sprawl.

Allen Scott portrays Los Angeles as the prototypical postmodern city.[11] This form of urbanism is generated by a structural change from what he calls Fordist to post-Fordist industrial organization. He contends that there have been two major stages of urbanization in the United States. The first was a period of mass production in the style of Ford Motor Company. Older industrial cities such as Detroit, Chicago, and Pittsburgh grew up around such forms of industrial organization. The second stage is associated with the decline of the Fordist era and the use of post-Fordist "flexible production," small-batch production units linked into clusters of economic activity. In Los Angeles, such clustering is manifest in labor-intensive craft forms—typically clothing and jewelry—and high technology, especially the defense and aerospace industries. As described by Michael Dear and Steven Flusty, such clustering has resulted in "a continuous collage of parcelized, consumption oriented landscapes devoid of conventional centers."[12]

The transition from Fordism to post-Fordism in Los Angeles has entailed substantial economic restructuring. Declines in automobile assembly and in furniture and plastics manufacturing have taken a heavy toll; cuts in federal defense and aerospace budgets have also significantly reduced productivity in these important sectors of the local economy. Although Los Angeles contained about half the statewide manufacturing jobs in 1970, its share declined to 41 percent of the state total in 1989 and to about 36 percent in 1995. Between 1983 and 2001, the unemployment rate in Los Angeles County fell from 9.7 percent to 5.7 percent—but during this period, 238,000 higher-paying manufacturing jobs were lost and 765,400 lower-paying service jobs were gained.[13] On the other hand, some growth can be noted in other economic activities, including entertainment, telecommunications, international trade, and biotechnology.

Because of its location at the edge of the Pacific Rim, Los Angeles is strategically located in East–West international trade, and the Los Angeles Customs District has become the busiest in the nation, surpassing New York in dollar volume of trade.[14] East–West trade is further enhanced by the diversity of ethnic groups in the city. According to the census of 2000, the ethnic makeup of the city is 47 percent white, 11 percent African-American, 19 percent Asian, and 26 percent other races; the population is 47 percent Hispanic of any race and 30 percent non-Hispanic whites. The Asian and Hispanic populations of the city have the language and cultural skills that are valuable for developing trade relations with Asian and Latin American business firms, the latter all the more so in light of the recently enacted North American Free Trade Agreement (NAFTA).

The rapid growth of foreign trade in Los Angeles has also been accompanied by the importation of huge quantities of foreign capital and labor. Edward Soja states that such factors have been as vital to the recovery and growth of Los Angeles as any domestic restructuring.[15] Most domestic high-rise developments and prime properties in the downtown are now owned by foreign capital, with investment from Japan, Canada, Britain, and the Netherlands. Surrounding the downtown, an inner residential ring has become the primary location of a largely Third World labor force.

> The city center is rimmed by an almost continuous girdle of ethnic neighborhoods: Little Tokyo, Koreatown, Little Manila, the great barrios of East Lost Angeles and Echo Park–Alvarado (home to most of 400,000 Salvadorans who have moved in since 1980), a downtown as well as suburban Chinatown, and an increasingly pinched salient of predominantly black Los Angeles. Except for skid row, still mainly black, the inner city is overwhelmingly Latino and Asian, drawn mostly from countries rimming the Pacific Ocean.[16]

In a chapter titled "It All Comes Together in Los Angeles," Soja contends that the city symbolizes the new urban paradigm because it is a fulcrum for three global trends: accelerated immigration, dispersed production, and the internationalization of the division of labor. In addition, Los Angeles performs an important global role: "A growing flow of finance, banking, and both corporate and public management, control, and decision-making functions have made Los Angeles the financial hub of the Western USA and (with Tokyo) the 'capital of capitals' in the Pacific Rim."[17]

But while these capital inflows have sustained the city's growth, the price has been deepening social and racial polarization. Capital inflows and increasing economic activity have put upward pressure on prices in the more desirable residential areas. Consequently, an ever smaller portion of the region's population can afford homes in those areas, and lower-income workers are forced into less desirable inner-city areas, increasing the polarization of housing prices in the city and causing social unrest. The Watts riots in 1965 reminded the country of the deep racial divisions that the city faced then. And in the spring of 1992, following the exoneration of police officers in the beating of black motorist Rodney King, some the worst rioting in U.S. urban history transformed Los Angeles into a war zone.

INNOVATION CENTERS

As part of the shift away from Fordism and toward more flexible forms of economic organization, a relatively new urban phenomenon has evolved, namely, the self-sustaining center of innovation that is linked into global markets. In these locations, continuous inputs of knowledge are far more important to the production process than they were in previous eras. Here, research and development are an ongoing activity, as short product cycles call for constant product improvement. Heavy reliance is placed on face-to-face and informal links between key entrepreneurs and innovators and the wider "innovative milieu" within which continuous innovation can flourish. Linkages to research institutes and universities tend to be important assets, in addition to good international transport and telecommunications. Key sectors include electronics and telematics, biotechnology, aerospace, nuclear technology, medical technologies, and environmental technologies.

As previously noted, important examples of such localized production complexes include the Silicon Valley in California, Route 128 in Massachusetts, and Research Triangle Park in North Carolina; other notable areas modeled after California's Silicon Valley are the Silicon Prairie in suburban Dallas, Silicon Mountain in metropolitan Denver, Silicon Forest between Portland and Seattle, and Silicon Desert near Phoenix (see box 4.4). Case profiles of the Silicon Valley and Route 128 are presented next.

Box 4.4. Selected Listing of Key High-Technology Centers

Austin, Texas	This area boasts Dell and about 2,000 other high-tech companies.
Boston	The area along Route 128 is the biggest recipient of venture capital after the San Jose area.
New York	Silicon Alley is a three-mile strip extending from Manhattan's trendy Chelsea district to its southern tip.
Salt Lake City	One neighborhood has more high-tech companies than fast-food restaurants.
San Jose, Calif.	Referred to as the Silicon Valley, this is the location of thousands of start-ups that developed Internet, telecommunications, medical, and pharmaceutical technologies.
Seattle	This Northwest city is the home of Microsoft and Amazon .com.
Washington, D.C.	The Northern Virginia suburbs' rich mix of high-tech firms and federal agencies make them a high-tech powerhouse with information technology–based employment comparable to San Jose.

Silicon Valley

The development of the Santa Clara Valley in California as a high-tech area can be traced back to World War II. Federal funding of weapons research at Stanford University in the 1940s initiated the formation of the microelectronics industry in the region. The invention of the solid-state transistor at AT&T's Bell Labs was a response to the military's demands for small, versatile electronic components.

The electronics-based companies that set up operations in the area in the 1940s and 1950s laid the groundwork for the location of the valley's first semiconductor companies. As the market for microelectronics expanded in the 1960s and 1970s, so too did the region. By making silicon chips, personal computers, and related goods, this former agricultural valley came to be known as Silicon Valley. Today, more than 70 percent of the manufacturing workforce in Santa Clara County is employed in high-tech industries, and many others are employed in occupations that service high-tech industries.

To account for this transformation, observers acknowledge the role played by Stanford University in facilitating innovation during the 1950s and 1960s. Located in Palo Alto, the university provided a supportive intellectual environment for many of the scientists and engineers engaged in experimental ventures. Stanford Industrial Park, located on 770 acres adjoining the university campus, was dedicated entirely to high-technology activities. Entrepreneurs could locate in the industrial park or, when space was not available, in the surrounding area in order to interface with other entrepreneurs. Following the Stanford model, nearby towns soon established their own industrial parks and offered various forms of incentives to attract technology-based companies.

By the late 1970s, growing pressure from international competition and the emergence of huge new markets were causing large-scale restructuring of the microelectronics industry. Most of the small companies were being bought up or merging with larger firms. In addition, almost all labor-intensive assembly operations were being relocated to low-wage areas in Asia and Mexico and, to a lesser extent, in Europe (in order to gain access to the European Common Market). Doomsayers were predicting the end of Silicon Valley and seemed to be vindicated when the rampant investment of the late 1970s and early 1980s dried up. However, new generations of small computers followed hard on one another, putting Apple's Macintosh and Sun workstations at the top of the charts by the end of the 1980s. All the while, Intel was building more and more circuitry on its central processing chips. As a result, Silicon Valley regrouped and became competitive once again.

A recession in the early 1990s cost many more people their jobs, but the Valley reconstructed itself again. Michael Lewis writes that the Valley continues to find new directions: "Actually building machines is today a slightly second-rate occupation: the computer has become a commodity. Factories are messy, workers unnecessary. These days, the thrill is dreaming up things for the ubiquitous computer to do. Software, in a word."[18] Artificial intelligence, special effects, virtual reality, and other fantastic concoctions of software all are arenas in which the Valley still leads. Because large corporations cannot control the rapid rate of change, the new game favors young engineers and computer scientists, crammed into office buildings nicknamed "incubators," where they work on developing new software technology (see box 4.5). Included among the Internet success stories are Netscape, America Online, Amazon.com, and Sun Microsystems.

Unfortunately, however, the industry remains quite volatile. In the economic crash of 2001, the region lost 200,000 jobs, and unemployment in the area hovered between 7 and 8 percent. The push for efficiency had sent jobs to less-expensive countries such as India and China. But, as in the past, the Valley's economic infrastructure—venture capitalists, lawyers, accountants, investment bankers, executive search firms, and an abundance of other specialized activities all centered on new startup firms dedicated to innovation—has made for a strong recovery.[19]

Box 4.5. Silicon Incubators

All across the Silicon Valley, office buildings are crammed with rogue inventors entrenched in facilities called "incubators." Incubators house engineers and computer specialists whose ideas may someday make a fortune but are currently too tenuous to survive in the open market. What do these people do?

Michael Lewis, who toured such a facility, describes one such person: a forty-nine-year-old researcher named Barbara Hayes-Roth, who has started her own company called Extempo. Experimenting in her Stanford laboratory, she succeeded in creating computer cartoon characters that could move about and respond to queries. With a little imagination, these characters were converted into electronic salesmen who can happily guide Internet consumers to the latest products of just about any manufacturer. In contrast to live salesmen, the electronic salesman always has a friendly smile, even for the most abusive customer.

In a neighboring office, a young man sits over his desk some fourteen hours a day, stroking his forehead, trying to invent new ways of using the phone. He wants to create software for an Internet telephone. No one knows how much progress he has made, because he refuses to discuss his ideas.

Across the hall from him, a twenty-seven-year-old Harvard Business School graduate sleeps under his desk. His name is Francis Tapon, and he and his colleagues formed Sightech Vision Systems to develop and promote software for a computer that can identify and remove the smallest defects in the manufacturing of a product. If all goes well, it is estimated that Sightech will be worth about $600 million.

Source: Michael Lewis, "The Little Creepy Crawlers Who Will Eat You in the Night," *New York Times Magazine*, 1 March 1998.

The Route 128 Phenomenon

Along with Silicon Valley, the Route 128 phenomenon in Massachusetts is a quintessential high-tech success story. Despite downturns in the early 1900s and 2001, new high-tech firms continue to spread along the famous Route 128 corridor around Boston, all the way to Worcester in the central part of the state and even into southern New Hampshire.

Although the area did not become nationally known until the 1980s, it was drawing modest attention as early as the 1950s. It was then that Ken Olsen founded Digital Equipment Corporation (DEC)—a company that more than any other came to symbolize the era. Olsen developed many of his ideas for a minicomputer while working as a researcher at Massachusetts Institute of Technology (MIT)–affiliated Lincoln Laboratory, which was being supported by federal funds. The evolution of the Route

128 region is strikingly similar to that of Silicon Valley. Most prominent in this regard are the combined effects of three powerful sectors: academia, the federal government, and industry. A major force behind the growth of Route 128's high-tech business firms has been the generation of innovative ideas that emerged from nearby universities. Though most engineering schools in the Boston area have established links with industry, MIT in particular has established a leadership role through its Industrial Liaison Program. The Boston area also has one of the largest academic communities in the nation, including Tufts, Boston, and Harvard universities.

As in California, the federal presence in Massachusetts during World War II served to catalyze technology in the Route 128 region. The development of advanced microwave technology, the construction of the first practical digital computer, and the creation of guidance systems for space flights are some of the Boston area's technological achievements that were supported by defense dollars. MIT's role was critical. MIT had already provided the foundation for technical talent, and wartime mobilization brought in a skilled workforce from government, industry, and other universities; it also provided the funds to introduce a whole new generation of technologies. When the war ended, many of the scientists and engineers who had been drawn to the area stayed to take advantage of the commercial marketplace. At the same time, federal research dollars continued to flow in and helped to position the region at the forefront of technical innovation.

Most of the high-tech firms in the Route 128 corridor started small and have tended to stay that way. This has served to reduce risk and to allow extra leeway for experimentation with ideas that could eventually be converted into new technologies. MIT's Technology Licensing Office reports that 85 percent of MIT patents have been licensed to small firms. It was small high-tech firms that led the way out of a severe recession in Massachusetts in the mid-1990s and again in the early 2000s. Even as software and Internet firms go bust, new ones are created in their stead.

OLD INDUSTRIAL CITIES

Data presented in chapter 2 demonstrate the extent to which the central cores of major urban areas have been losing many of their traditional functions such as manufacturing, retailing, and wholesaling. Most of these are older cities in the Northeast and the Midwest and, as such, they are subject to considerable stress. Using a technique to assess urban hardship, the Rockefeller Institute developed the Intercity Hardship Index, which compares the economic condition of U.S. cities.[20] With data from the 2000 Census, the index for each city is derived from six key factors:

- percentage of unemployed
- percentage of the population defined as dependent
- percentage of the population with less than a high school education
- per capita income
- crowded housing
- percentage of people living below the federal poverty level

Those cities that appear to be most troubled are listed in table 4.3.

With some exceptions, as in the cases of Los Angeles and Miami, such cities tend to manifest two disturbing qualities: They are dependent on control centers located elsewhere, and they are expendable in the system of places. That is to say, they do not have any of the distinctive organizational features of headquarters or of innovation centers that would allow them to control their own destiny. The functions they perform can be carried out in any number of other places. For example, old factory or port sites of the Northeast and Midwest such as Gary, Cleveland, Detroit, Hartford, Norfolk, and Newark, among others, have been especially vulnerable to mergers and downsizing decisions made by distant companies with global interests. The efforts of leaders in these cities to reverse course by remaking the physical and economic infrastructure has so far produced mixed results at best. The case histories of Detroit and Newark provide illustration.

Detroit

As the world capital of automobile production, Detroit is known as the Motor City. Yet the past score years have been hard on this city of just

Table 4.3.   Cities with the Most Hardship in 2000

City	Region	Hardship Index Score
Santa Ana, Calif.	West	73.7
Miami	South	71.6
Hartford	Northeast	67.1
Newark	Northeast	66.6
Gary, Ind.	Midwest	59.4
Detroit	Midwest	56.6
Cleveland	Midwest	55.8
Fresno, Calif.	West	54.4
Los Angeles	West	51.0
Buffalo	Northeast	50.1

Source: Lisa M. Montiel, Richard P. Nathan, and David J. Wright, "An Update on Urban Hardship" (Albany, N.Y.: Nelson A. Rockefeller Institute of Government, 2004), 4.

under one million people, which has been losing automobile manufacturing jobs to Japanese, European, and Korean companies. In fact, only
three automobile assembly plants are left in Detroit. The city continues
to lose population and what remains are predominantly poor blacks. According to the 2000 U.S. Census, 82 percent of the population is African-
American and 5 percent is Hispanic. White flight has become "bright
flight," with families and people earning more than $50,000 a year lead-
ing the way out of town.

While the suburbs continue to boom with new service industry jobs, Detroit has lost some 15,000 businesses since 1972, and the unemployment rate
hovers around 14 percent. To deal with the poverty rate, which is around 33
percent (the second highest in the nation), the public sector has become the
top employer, the health care industry second, and the auto industry third.
However, with the city facing multimillion-dollar shortfalls in revenue collection and the threat of receivership, Detroit has been forced to lay off
workers and cut employees' pay 10 percent across the board, in addition to
ending overnight bus service and closing the aquarium. The struggle to attract businesses and residents has been compounded by the city's tax bur-
den which is 5.5 times that of the average municipality in Michigan.

Detroit can be described as a mix of spot renewal amid urban ruin,
which dates back to the rioting and firebombing of the late 1960s. Rod
Gurwitt provides an apt description:

> Rising from the downtown streets are blocks-full of graceful 1920s skyscrap-
> ers ornamented with intricately carved griffins, Indian chiefs and explorers,
> their charm sabotaged by boarded facades and broken upper-story windows.
> One can drive through decayed residential areas, then pass suddenly into a
> neighborhood of immaculate lawns and large, gracious homes.[21]

Racial tensions are not far below the surface. For years, suburban bus
routes were designed to avoid matching up too closely with city routes
near the Detroit line, making it difficult for mostly black city residents to
reach the mostly white suburbs. Moreover, the city has no subway or commuter rail lines in its 140 square miles, and it ranks last in public transportation among the nation's twenty largest metropolitan areas. So while
suburbanites have no trouble driving to jobs downtown, city dwellers
without cars struggle to get to the jobs that would lift them from the welfare or unemployment rolls.

Though the city acquired an empowerment zone where tax and other
incentives are available to promote development, the rebuilding process
has been problematic at best. In the late 1970s, to jump-start the economy,
a consortium of more than fifty corporate investors led by the Ford Motor
Company laid out $350 million to build the Renaissance Center—a com-
plex of office towers, an atrium, and a seventy-six-story hotel that fronts

the Detroit River. This project, together with a $2.9 million monorail that was completed in 1989, was expected to spur development in the downtown business district. However, many of the tenants who moved to the towers simply abandoned other buildings, making it a zero-sum game. In 1997, General Motors purchased the Renaissance Center to serve as headquarters for its 10,000 employees—down from 126,000 workers in the city in 1984. The fact that the complex sold for less than a quarter of what it cost to build two decades earlier reflected the real estate market's wariness in investing in a deteriorated downtown.

Though the Detroit metropolitan region continues to be the center of America's automobile industry, much of the industry has been decentralized since the 1970s. For example, General Motors established an automobile design center in Southern California to generate new ideas for the building of its cars. Similarly, the growing practice of joint production between U.S. and foreign firms means that "headquarters is less of a headquarters."[22] General Motors' Saturn cars, although made in the United States, are based on Japanese designs, and the more recent mega-merger of Germany's Daimler-Benz with Chrysler has shifted control away from Detroit.

Public–private partnerships to invest in workers and businesses are key factors in strategies to achieve an economic comeback. For example, Ford, GM, and Chrysler have targeted procurement to minority-owned and empowerment zone–located supplier firms; a Detroit Investment Fund has been organized by corporate leaders to stimulate business start-ups and to attract companies to the city; Detroit-area financial institutions have made commercial and industrial loans to businesses located in the empowerment zone; a community development bank has been opened; and a Detroit Works empowerment training partnership with the Carpenters Union and Painters Union prepares empowerment zone residents for construction jobs.

Giving some hope for a comeback are a wide range of new developments, including a new downtown baseball stadium for the Detroit Tigers and a new football stadium nearby for the Lions (which moved back to the city after playing for over a decade in the suburban Silverdome). New cultural and entertainment facilities, including a new opera house and three gambling casinos, have been built downtown.

Newark

Not too long ago, Newark—New Jersey's largest city—hummed with breweries, tanneries, shipyards, and factories turning out shoes, thread, jewelry, paint, military supplies, and radio equipment. But when manufacturing no longer required proximity to markets, investment capital began to drain out of Newark, leaving behind empty factories and stores, un-known quantities of toxic wastes, and thousands of unemployed workers. A devastating riot in 1967, which left twenty-six people dead and much of the city in ruins, fomented further disinvestment. Between 1970 and the late 1980s, the population of Newark decreased 27 percent—from 382,417 to 275,221 (at last count in 2000, the population was stabilized at 273,546). In addition, the city lost a fifth of its total assessed property value and, with declining revenue, was compelled to eliminate more than a quarter of its municipal workforce.

What has remained is one of the poorest of the nation's big cities, with a third of its residents living below federal poverty standards and the un-employment rate at around 12 percent. Fifteen percent of people between the ages of sixteen and nineteen are high school dropouts. According to the 2000 Census, furthermore, nearly 60 percent of the population is African-American, approximately 30 percent is Hispanic or Latino, and only 18 percent is white.

Though unemployment remains high, the nonprofit New Community Corporation formed in 1968 has placed thousands of low-income resi-dents in jobs with private employers throughout the metropolitan region. The New Community Corporation also runs a large child care center and has built or renovated more than twelve thousand units of affordable housing. It is co-owner with Pathmark Stores of a seemingly successful shopping center, the first new retail facility built in Newark's Central Ward since the 1960s. These efforts are being supplemented by an ambi-tious program to transform Newark's large public housing stock. The city is currently demolishing thousands of units of deteriorated, vacant public housing in high-rise buildings and replacing those units with attractive low-rise townhouses for rent and ownership. Recent efforts at revitaliza-tion under the administration of Mayor Sharpe James have included the 1997 opening of the New Jersey Performing Arts Center and a minor league baseball stadium. The latest major development plan for down-town is a $355 million arena for the New Jersey Devils hockey team.

Recognizing improvements, the National League of Cities presented Newark with an All American Cities award in 1993. But if "the phoenix is rising," as Mayor James proclaimed, its flight has not reached the neigh-borhoods that have lost thousands of residents and where poverty con-tinues to plague the Newark as a "dual city" where most the population are the have-nots. The city's schools system is considered one of the worst in the state, and the city itself is second only to New York City in its per capita rate of AIDS. According to the 2002–2003 Newark Kids Count sur-vey, Newark's children are less likely to receive immunizations, more likely to fail in school, and more likely to suffer from health problems than children living elsewhere in New Jersey.

BORDER CITIES

Another type of city, as described by Logan and Molotch, is the U.S. border city, which performs a unique economic function.[23] Broadly viewed, these cities are typically labor centers consisting of a large immigrant population with cultural and economic ties to places south of the border—primarily Mexico, Central America, and Cuba. Many of these workers are undocumented (without proper papers), which makes them vulnerable to exploitation in the form of low wages and poor working conditions. For the most part unskilled, they work as busboys, waiters, and maids in tourist areas. Others work as landscape day-laborers, field hands, building cleaners, and factory worker, often under sweatshop conditions. Employers see them as a controllable labor force, more productive and less costly than resident workers.

One U.S. city that fits this profile is Tucson, Arizona, where approximately one-third of the population is Hispanic. Box 4.6 conveys some of the social dynamics that are fairly typical of such places. Other cities are Miami, where two-thirds of the population is Hispanic; San Antonio, where over 50 percent of the population is Spanish speaking; and El Paso and Brownsville, Texas, where more than 75 percent of the population is Hispanic. To a certain extent, such large California cities as Los Angeles, San Diego, and San Francisco, with large Hispanic and Asian populations, have border-city features in addition to their other roles.

Border cities function as trade and financial centers, importing, marketing, and distributing goods. NAFTA, which eliminated trade barriers between the United States, Canada, and Mexico, greatly expanded the role of border cities as trade and distribution centers where legal and illegal goods are transported across the border. Since 1994, when the trade agreement went into effect, a growing number of U.S. manufacturers have been inclined to take advantage of low wage rates in Mexico, while maintaining offices in border cities. Drugs and other illicit monies from Latin America can be laundered through the growing banking industry on the other side.

Contributing to a growing concern over illegal border crossings has been the fear of terrorist infiltration, which followed the terrorist attacks on the World Trade Center and the Pentagon in 2001. Despite an influx of new technology, such as underground sensors and cameras that span the desert, Border Patrol agents catch only about one-third of the estimated three million people who cross the border illegally every year.

We also see that the very rich of Latin America do their shopping in cities like San Diego, Los Angeles, Dallas, and other border locations. They use these places as secure havens for their wealth, as well as for recreation and culture. "All of this is made possible, of course, by U.S.

Box 4.6.   Tucson North and South

Soon after World War II, thanks largely to air-conditioning and improvements in highway and air transportation, Tucson experienced a major land boom. From a population of 35,000 in 1940, the city grew to 212,000 in 1960. The communications revolution of the 1980s and 1990s, combined with the appeal of Tucson's warm and scenic desert attributes, contributed even more growth—roughly 465,000 residents at last count. In the Catalina foothills in North Tucson, a relatively new cohort of upper-class residents using the latest in electronic communications conducts business with clientele in distant places of the country and around the world. Other new arrivals to the city are mostly wealthy retirees.

Because the local economy has not kept pace with the growth of population, there are now two Tucsons, one on the north side of town and the other on the south side. With the exception of the University of Arizona—"somewhat of a social and economic island"—and the economically erratic military aircraft industry, there is little available employment other than low-paying service jobs. North Tucson is largely white, while South Tucson is primarily Mexican, comprising both legal and illegal immigrants. The business section of the south side consists primarily of auto repair stores, tire shops, and Mexican restaurants. According to one observer (a retired Tucson policeman), "Tire shops have traditionally been fronts for drug deals or for laundering illegal cash. . . . Arrest all the drug dealers and the retail economy of Tucson's South Side goes bust." Interviews reveal that Southside Mexicans simultaneously hate and envy North Tucson. According to a "semi-retired" local gang leader, they have no understanding of how to save and invest money to achieve upward mobility. What they do know is, "If I could only sell a bunch of keys [kilos of cocaine], I could move to North Tucson."

Source: Robert K. Kaplan, "Travels into America's Future," *Atlantic Monthly* 282 (July 1998): 56, 58.

laws and administrative procedures in immigration policy, foreign trade, banking, and product copyright, which funnel a massive stream of goods and people through these specialized gates of regulation and evasion."[24] Consider the case of Miami, Florida.

Miami

Miami, a city of 362,000 people, is one of twenty-eight municipalities in Dade County, Florida, which has a population of over two million and recently changed its name to Miami-Dade County. On any number of criteria—immigration, trade, finance—Miami-Dade County is one of the most internationalized metropolitan areas in the nation. This was not always so. Some forty years ago, it was a quiet resort area at the far end of the Florida peninsula. Today it is a growing metropolitan region strategically positioned between North and South America and the Caribbean. From 1960 to 2000, the number of people living in Dade County more than doubled, and by 2000 its economy was worth an estimated $60 billion, eclipsing the output of some nations.

Miami's transformation is based on two interrelated developments: the immigration of large numbers of Latin Americans and the globalization of the world economy. Today, Latinos, approximately 66 percent of whom are Cuban, constitute more than half the region's two million people. The presence of large numbers of bilingual Spanish-speaking residents makes Miami an inviting location for firms that look to do business in Latin America. More than a third of all U.S. trade with Latin America and over half of all U.S. trade with countries in the Caribbean and Central America flows through the region. Of the many multinational companies with offices in Miami, 70 percent were established after 1980, and the number of new establishments has continued to grow. In addition, Miami has become the third busiest foreign banking center in the United States (after New York and Los Angeles), as measured by the number of foreign banking offices. And while tourism remains an important source of income, it is increasingly focused on international markets, especially in countries in the Southern Hemisphere. Thus, the Greater Miami region has become the leading gateway to Latin America and the Caribbean, beating out such cities as Houston, Los Angeles, New York, and New Orleans.

But if Miami has emerged as a world-class trading center, the city and the region are in many ways constrained by a lack of social cohesion and by economic polarization. According to the U.S. Census Bureau, Miami had more residents living in poverty in 2000 than any other U.S. city. A report from the Brookings Institution indicates that both Miami and Miami-Dade County have only a small middle class and that the disparities between rich and poor are growing.[25] Furthermore, blacks and Hispanics are less likely to be in the middle class than whites are, as white median household income is at least $20,000 more than the black, Puerto Rican, Nicaraguan, and Haitian median household incomes. A contributing factor is that the regional economy is a low-wage economy: Most jobs are in industry sectors such as service and retail that pay lower wages.

Adding to the region's problems is an inept, often perverse system of government. Nearly every branch of local government has been sullied by scandal. A wide-ranging federal investigation into corruption in city and county government in 1996 and 1997 resulted in the conviction and imprisonment of a former city commissioner and a former city manager, and the mayoral election of November 1997 was invalidated by the courts because of fraud. In addition, officials associated with the Port of

Miami, including its ousted former director, were convicted of stealing $1.3 million from the agency. On top of these incidents of official corruption, Miami struggles with budget deficits to the point that it teeters on the edge of bankruptcy. Consequently, then governor Lawton Chiles appointed a special board to oversee the city's money.

RETIREMENT SITES

A demographic trend of growing significance in the United States is the aging of the population and the tendency toward early retirement. According to the U.S. Bureau of the Census, persons sixty-five years of age or older numbered 35.9 million in 2000. They represented 12.3 percent of the U.S. population and one of every eight Americans. By 2030, it is projected that there will be 71.5 million persons in this age bracket, representing 20 percent of the populations.

An important aspect of retirement for most people is deciding where to live. In the past, this was not an issue, because most elderly persons remained at home where they could be close to family. In contemporary times, parents and grown children often go their separate ways, staying in touch by telephone, or more recently through cell phones and e-mail. Every year, more than 400,000 adults fifty-five or older move out of their home states and relocate. Florida leads all states in the proportion of elderly residents—20 percent over age 65—most of whom have relocated from other places. Among those places prominently mentioned are Boca Raton, Daytona Beach, Fort Lauderdale, Fort Myers, Saint Petersburg, and Ocala, where close to 20 percent of the residents are over sixty-five years of age. Sarasota has an even higher proportion of 32 percent. Additional cities with reputations as retirement centers are Savannah, Georgia; Prescott and Scottsdale, Arizona; Palm Springs, California; and Asheville, North Carolina. Even tourist centers like Las Vegas and Reno, Nevada, where residents pay no income tax and no inheritance or estate taxes, are attracting growing numbers of seniors. Many smaller, private development communities are packaged as state-of-the-art retirement communities, offering such amenities as golf courses, lakes for fishing and boating, and cultural and recreation activities. They go by such names as Sun City Grand near Phoenix; Sun City Palm Desert near Palm Springs, California; Sun City Summerlin near Las Vegas; and Ford's Colony close to Williamsburg, Virginia, among others.

Some states, mostly in the South, and hundreds of towns and cities have been actively recruiting young retirees as a way of enlarging their revenue base. This includes many college towns that have targeted seniors by offering cultural events and reduced fees for college courses. Cities such as Ithaca, New York (the home of Cornell University); Bloomington, Indiana (Indiana University); Hanover, New Hampshire (Dartmouth College); and Blacksburg, Virginia (Virginia Tech) have been offering special incentives to encourage private development of retirement housing.

To the extent that the proportion of retired elderly people in towns and regions continues to grow, certain consequences tend to follow. When such communities are essentially residential, their revenue base depends largely on the individual wealth of their residents. Indeed, the economies of such communities are becoming increasingly dependent on the rise and fall of pensions, Social Security, and Medicare payments. Charles Longino, the author of *Retirement Migration in America*, states that retirees are also likely to block future industrial development efforts since their priorities emphasize nice climates and attractive places to live and recreate.[26] They are inclined to reject the noise and congestion that come with new factories and the jobs they bring to the economy.

In her book *Young v. Old: Generational Combat in the 21st Century*, Susan A. MacManus notes that the higher turnout rates among older voters in state and local elections tend to affect funding for programs that directly benefit one age group over another.[27] This is most obvious in funding battles between younger families who favor education and older people who place greater importance on law enforcement (see box 4.7). Consider the case of Saint Petersburg.

Saint Petersburg, Florida

Saint Petersburg, population 248,000, lies on Tampa Bay on Florida's west coast. Called the Sunshine City because of its pleasant climate, it has become virtually synonymous with retirement and old age. People sixty-five years old and over make up about 17 percent of the city's population. Saint Petersburg and Tampa, which lie across the bay from each other, form part of a metropolitan area that has more than two million residents.

The city's origins as a haven for elderly persons date back to 1885 when, at the annual meeting of the American Medical Association, Dr. W. C. Van Bibber of Baltimore delivered the results of his search for the ideal location for a "World Health City." He selected Pinellas Point, Florida, located on a small peninsula in the Gulf of Mexico. This pronouncement from a medical authority created interest in the area, and in 1895, a Philadelphia entrepreneur managed to entice a large number of hopeful individuals to relocate to the recently incorporated village of Saint Petersburg. Many of these individuals were wealthy, and in its early years, the village was thriving. Over time, as Saint Petersburg experienced the ups and downs of various land development and complicated business schemes, the city

---

Box 4.7.   Welcome to Eldertown

The retirees don't necessarily try to put a bunch of people on schools boards who would run the schools cheaply. The kicker lies elsewhere. When it comes to pass a bond for a new school or arts center, the retirement community folks seem to vote the issues down handily.

During the 1970s, when Sun City became a major presence in Arizona's Peoria Unified School District, eighteen school bonds in a row were defeated by the Sun City residents; all other precincts within the school district voted to approve the bonds and the rise in property taxes it would mean. For young families, the problem was acute: Peoria schools were so crowded that the school system was holding double sessions, using church basements for some classes and even, in one school, a cleared-out janitor's closet for a special education class of four children.

Eventually, Peoria solved its problem by letting Sun City secede from the district. It meant Sun City residents didn't have to pay any school taxes, and that made them more than willing to opt out. For its part, the school district lost 75 percent of its assessed valuation. But says Bill Maas, the school district's assistant superintendent and a third-grade teacher at the time of the secession, "We lost a lot of our tax base—we went from $135 million to $35 million—but a lot of families have moved here. We're up to $500 million now. The schools are in good shape. We have never failed to pass a bond election since Sun City seceded."

Source: Reprinted by permission of the publisher from Penelope Lemow, "Welcome to Eldertown," *Governing* 10 (October 1996): 23.

---

continued to be viewed as a Shangri-La for the elderly and the physically frail.

A real estate boom took place in 1912, and in 1914 the first commercial airline began flying people across the bay that separated Saint Petersburg from Tampa. Several major league baseball teams held spring training in the city from 1914 on, and as a result the city got a reputation as the Winter Baseball Capital of the United States. The city's first major league baseball team, the Tampa Bay Devil Rays, began playing in Tropicana Field in 1998. Because it lacked a navigable harbor and natural resources that helped Tampa develop as a hub of commerce and industry, the Sunshine City evolved as a place of tourism and suburban development.

Early settlement tended to be concentrated along the bay. This section became the original downtown, where many of the old tourist and residential hotels remain. As Saint Petersburg developed outward during the 1960s and 1970s, young families and wealthy retirees built homes further from the downtown. At the same time, tourists were finding the beaches along the Gulf Coast more attractive than the wide verandas and formal dining rooms of Tampa Bay hotels, and suburban malls were soon replacing stores in the central business district as places for shopping. Out of economic necessity, many large homes built in the 1920s had to be divided and subdivided, creating a low-rent district. As a consequence, a pattern of age segregation emerged for retirees of modest means.

The greatest concentration of elderly persons continues to be in the old downtown neighborhoods, where social clubs, thrift shops, cafeterias, and retail stores for the elderly are also located, as is the city's multiservice senior center, built in the late 1970s.

Historically, attempts by the business community to change the city's image as a place primarily for the elderly has antagonized the older downtown residents. For example, in 1961 the City Council voted to remove several thousand green benches from the city sidewalks—benches which were regularly used by older residents and which they identified as their trademark. As described by Maria D. Vesperi in her book *City of Green Benches*, "This seemingly minor issue marked the emergence of a social and economic engagement that has troubled St. Petersburg ever since."[28] When the municipal Pier Building on the city's waterfront was demolished in 1967 to be replaced by a modern structure shaped like an inverted pyramid, older people viewed the new Pier as unattractive and another purge at their expense. Meanwhile, Saint Petersburg, keenly aware of how neighboring Tampa was progressing (with new center city skyscrapers, a fancy airport, a professional football team complete with stadium), was moving ahead with its own modernization, constructing a new yacht basin, refurbishing its Bayfront Center, and rebuilding the Pier again in 1988.

If relations with the elderly have been strained, there have also been problems with African Americans, who comprise a little more than 20 percent of the city's population. Many of Saint Petersburg's oldest black residents arrived early in the twentieth century from rural Alabama, Georgia, and South Carolina. Women were usually employed in the hotels, restaurants, and private homes, while the men worked in construction and on the railroads. During the 1960s and 1970s, as the regional economy was changing, the poverty rate among blacks grew, reaching about 25 percent in 2000—three times the white poverty rate. Racial unrest erupted in the fall of 1996 after a white police officer fatally shot a black teenager in a traffic stop, and again a short time later when a grand jury ruled the shooting justified. Although city officials say blacks have reached representative numbers on city boards and commissions and in the rank-and-file of the police department, economic inequities and complaints about police treatment have persisted.

LEISURE-TOURISM PLAYGROUNDS

As reported by the Travel Industry Association of America's *Tourism Works for America* (2002), travel and tourism in America generate more than $568 billion a year in spending and employ about 8 million people directly and 9.4 million people indirectly.[29] Approximately $98 billion is generated in tax revenue for federal, state, and local governments. International travelers spend $91.1 billion in the United States. By these or any other measure, tourism and the pursuit of leisure activities constitute one of the largest industries in the United States, playing a major role in shaping the urban landscape.

Strategies that cities have adopted for capturing tourist dollars include promoting theme parks, gambling casinos, museums, and consumer shopping (see box 4.8). Places such as Las Vegas, Nevada, which lack a long historical past, have created a tourism infrastructure from the ground up based on gambling and theme park fantasy derived from Disney World and Disneyland. Old industrial cities that have experienced decline have come to rely, more or less, on images of their historical pasts. For example, the city of Bethlehem, Pennsylvania, has converted its old steel mills into a museum to educate visitors on what the industrial era looked like. Other cities have focused on professional sports.

In her book *The Unreal America: Architecture and Illusion*, Ada Louise Huxtable describes these trends as representing a fundamental change in the way the general public perceives and understands communities: Increasingly, architecture and the urban environment, as they are being designed, have come to symbolize disengagement from communities as they really are, with the idea of making them into a more agreeable product.[30] As she explains:

> Surrogate experience and synthetic settings have become the preferred American way of life. Environment is entertainment and artifice; it is the theme park with the enormously profitable real-estate bottom line and a stunning record as the country's biggest growth industry.[31]

Propelled by new technical devices—robotics, online videos, computer animations, virtual reality—fantasy environments are now manifest not only in theme parks but also in historic restorations, resort areas, gaming complexes, museums, and shopping malls. Two case examples have set the tone in this country: Walt Disney Company theme parks and the gambling casinos of Las Vegas. As part of this general trend, furthermore, city centers are being redesigned as entertainment zones for tourists with the intent of revitalizing them. Downtowns that just twenty years ago more closely resembled ghost towns in Cleveland, Baltimore, St. Louis, and

Box 4.8.   Key Events in the Evolution of the Leisure-Tourism City

1955	Disneyland opens in Anaheim, California.
1969	The state of Nevada permits corporate ownership of gambling casinos.
1971	The Magic Kingdom of Disney World opens near Orlando, Florida.
1981	Faneuil Hall opens in Boston, beginning the festival marketplace phenomenon that soon spreads to other cities.
1982	Disney's EPCOT theme park opens at Disney World, exhibiting futuristic prototypes.
1985	The West Edmonton Mall is constructed, the first shopping center to combine shopping and entertainment.
1989	Disney–MGM Studios theme park opens at Disney World.
1991	The first Planet Hollywood theme restaurant opens in New York.
1992	Oriole Park at Camden Yards opens in Baltimore, Maryland, the first of a new generation of traditional-type ballparks located in central cities.
1993	Universal CityWalk opens in Hollywood, California.
1996	Niketown opens in Manhattan, representing a new form of retail theater.
1997	Disney opens the renovated New Amsterdam Theater in midtown Manhattan as a major strategy for rejuvenating Times Square.
2000	The Detroit Tigers baseball team decides to stay in center city as they move from Tiger Stadium to their new Comerica Park.

Source: Adapted from John Hannigan, *Fantasy City: Pleasure and Profit in the Postmodern Metropolis* (New York: Routledge, 1998).

Washington, D.C., have become tourist meccas thanks to sports arenas and exhibition centers. Universal Studio's new CityWalk mall in Los Angeles and the restructuring of New York's Times Square have garnered special attention as high-tech prototypes (see box 4.9).

Disney Theme Park Centers

No one knows for sure when it all started, but certainly the Walt Disney Company has played a key role in redesigning much of the American cityscape into what John M. Findlay calls "magic lands."[32] The company operates theme parks and resorts, produces and distributes motion pictures, and runs a cable television channel. It also sells publications, videocassettes, videodiscs, and merchandise based on the Disney name and characters. In 1996, the company acquired Capital Cities/ABC, the broadcasting

---

Box 4.9. Times Square Comes Alive

Long recognized as the "Crossroads of the Nation," Times Square has once again become a place of bright lights and grand entertainment. For the better part of a century, this open space at Seventh Avenue and 42nd Street in Manhattan has served as a magnet for tourists from all over the world. Unfortunately, during the 1970s and 1980s, it fell on bad times. Despite its many theaters and fine restaurants, it grew increasingly squalid and crime-ridden. Tourists used the square as a place to pass through rather than to stay and enjoy.

Then, beginning in the early 1990s, through public and private investment, the Times Square area began to come alive once again. MTV, the rock-'n'-roll network, and Condé Nast, the trendy publisher of *Vogue* and *Vanity Fair*, set up shop here. Disney came in to restore an old theater and produce the long-running show, *The Lion King*. New office buildings, new restaurants, new shops and arcades, other newly restored theaters, and eye-popping electronics soon followed to make Times Square the Great White Way all over again.

Source: Doug Stewart, "Times Square Reborn," *Smithsonian* 28, no. 11 (February 1998): 36–37.

---

and publishing company that owns the ABC television network. Walt Disney also owns and operates Disneyland in Anaheim, California, which opened in 1955, and Walt Disney World (see box 4.10), whose Magic Kingdom opened in 1971 near Orlando, Florida. Disney World grew when EPCOT (the Experimental Prototype Community of Tomorrow), which features futuristic technology and cultural exhibits, was added in 1982. In 1989 Disney World added Disney–MGM Studios Theme Park, a movie studio with exhibits and shows. Subscribing to the belief that "more is better," in 1997 the company built a 500-acre "safari park" with a thousand wild animals that tourists can view from the safety of special safari vehicles. Furthermore, the company's international operations include Tokyo Disneyland, which opened in 1983, and Disneyland Paris, which opened in 1992, and Disneyland Hong Kong, which opened in 2005.

Underlying these theme park ventures are real estate deals and large-scale development projects. In 1984, the Walt Disney Company formed the Disney Development Company. As described in promotional material, its mission is "to plan, develop and operate real estate and new business opportunities compatible with Disney's entertainment mission." Ada Louise Huxtable elaborates:

> The theme park as focus and opportunity for land development is one of the most lucrative of all investments. Ostensibly (and practically) the surrounding land is bought and held by the park's owners for expansion, including its own hotels and resorts. But the big payoff comes with office complexes,

---

Box 4.10.   Disneyland: Animating the American Vision

Perhaps the most subtle seduction of Walt's grand theme park came from its evocation of national values. In myriad ways that visitors encountered consistently but perceived only half consciously, the park offered a remarkable distillation and reaffirmation of postwar American culture. The hordes of middle-class families who streamed into Anaheim in the decade after 1955 found themselves completely submerged in a fantasized but nearly pitch-perfect representation of their deepest commitments and beliefs. . . .

As visitors moved through Disneyland in the years that followed they encountered a series of populist political emblems that further reinforced an American way of life. The park promoted an unproblematic celebration of the American people and their experience. Main Street USA, with its nostalgic images of turn-of-the century small-town life, the heroic conquest of the West represented in Frontierland, the sturdiness of the heartland reflected in the Rivers of America, the Jungle Cruise in Adventureland with its playful pacification of the Third World, the promise of continued technological progress with Monsanto's House of the Future in Tomorrowland. The showcasing of sophisticated robot technology in the early 1960s—Audio-Animatronics, in Disney parlance—enhanced Disneyland's celebration of the American people. The Enchanted Tiki Room, which initiated the technology at the park in 1963, created a jovial melting pot atmosphere as its brightly colored electronic birds comically represented French, German, Irish, and other stereotypes. But Great Moments with Mr. Lincoln was probably the culmination of the park's roboticized version of American values. In this attraction, an electronically controlled replica of the sixteenth president rose to his feet against a swelling backdrop of patriotic music and solemnly paid homage to the tradition of democratic constitutionalism in the United States.

Source: Copyright 1997 by Steven Watts. Reprinted by permission of Houghton Mifflin Company. From Steven Watts, *The Magic Kingdom: Walt Disney and the American Way of Life* (Boston: Houghton Mifflin, 1997), 392–93.

---

malls, housing developments, and other commercial construction owned, operated, sold or leased, leased back, retained or disposed of by any number of elaborate real estate mechanisms . . . no one does it better than Disney.[33]

And while Disney gets bigger, so does everything else in the surrounding environment. With more than fifty million visitors a year and high hotel occupancy rates (over 80 percent), Orlando is already one of the country's top tourist destinations, along with Las Vegas and New York. Orlando International Airport has undergone a $1.2 billion expansion, including a new terminal complex and a new satellite for the existing terminal complex. Convention business is booming and a new convention center is planned in neighboring Osceola County.

Las Vegas

Over the last twenty-five years or so, increasing numbers of communities have seized on a range of seemingly ingenious gaming devices—from lotteries to off-track betting—that promise quick and painless ways to generate revenue. Legalized gambling casinos, first introduced in Las Vegas, Nevada, in the 1930s, are viewed as a sort of magic bullet that can spur growth and lower unemployment. To this day, Las Vegas is the classic example of a city that has developed gambling as a major industry. As such, the city is constantly reinventing itself to maintain its lead in an increasingly competitive economy.

As the fastest growing metropolitan area in the nation, Las Vegas has been booming. For every 100,000 people who come to the area to play, 250 come to stay. That adds up to 75,000 new residents a year; but despite the rush of newcomers, unemployment remains low. There is no corporate or personal income tax, so families that can't afford Phoenix or Los Angeles are inclined to settle in Las Vegas. A few years ago planners predicted that the Las Vegas Valley would be home to one million people by 2000. That milestone was reached in 1994, and two million are now being predicted by 2010.

This wasn't always so. In 1930, the year before gambling arrived and construction on Hoover Dam began, only 5,165 people inhabited the sun-dried valley. A moderate spurt in the economy occurred when nearby Hoover Dam imported about 5,000 workers, who patronized the hotels, bars, and most notably, the casinos legalized by the state in 1931. Many of them subsequently stayed to work at nearby military installations during World War II.

Conditions changed dramatically immediately following the war. There was only one luxury hotel out on the Strip when Bugsy Siegel, a New York gangster, arrived in 1946 to build his elegant Flamingo Hotel. Though Bugsy was promptly murdered by business associates, the next ten years witnessed the rapid proliferation of casino hotels along the Strip and downtown—almost all of them with underworld connections.

In 1966, the mysterious billionaire Howard Hughes arrived. Though he remained holed-up and isolated in his hotel suite for nine years, he engaged in casino shopping, buying the Desert Inn, the Sands, the Frontier, the Castaways, the Silver Slipper, and the Landmark. Following Hughes, MGM, Hilton, and Holiday Inn quickly became casino owners, too.

Casino gambling remains the economic backbone of Las Vegas: Thirty percent of all jobs are in hotels, gaming, and recreation, and those businesses in turn support most of the other service and construction workers. But because the ring of slot machines can now be heard in other cities, as well as on Native American reservations across the country, Las Vegas has

tried hard to market itself to a wider clientele. In the 1990s, it began to focus on family theme park–style entertainment. Bright lights, once the famous Strip's trademark, no longer sufficed as a prime attraction. The city
was forced to become more sensational, erecting a fake volcano that spouts
flames into the night; re-creating eighteenth-century sea battles across a
massive concrete lagoon; importing dolphins that frolic and even give birth
in a hotel pool; and devising sidewalk spectacles like the Excalibur's fire-
belching dragon. Atlantis is one of four extravaganzas at Caesar's; a computerized Bacchus and friends engage in a sound-and-light spectacular on
the half-hour, and in one rotunda the sky goes through a twenty-four-hour
cycle every hour. The exterior of the New York New York casino-hotel features a forty-seven-story reproduction of the Empire State Building, a 300-
foot-long copy of the Brooklyn Bridge, and a 150-foot replica of the Statue
of Liberty; a Coney Island roller coaster circles the whole complex. Not to
be outdone by the Strip, each night at dusk, the downtown area showcases
a sound-and-light show on a canopy over Fremont Street, featuring more
than two million lights and 540,000 watts of sound. Dancing and moving
images, some filmed, some animated, swirl above the street.

However, there is evidence that Las Vegas is in the process of reinventing itself once again. This is because the number of visitors who gamble
appears to be declining and the amount of money that the average gam-
bler is willing to lose is also declining. As one observer has noted, part of
the problem is that the market for gamblers coming in on Thursday and
flying out on Sunday is static.[34]

Consequently the new game in town is real estate development. Booming real estate prices have been the driving force behind the creation of a
dense, mixed-use mini-city as opposed to simply another large hotel and
casino complex. Most impressive is a project proposed by MGM (owner
of the MGM Grand and Mirage hotels) called City Center, which will be
the largest privately financed development project in the nation, covering
sixty-six acres near the Strip. The first phase of City Center is to include a
4,000-room hotel-casino; three 400-room boutique hotels; about 550,000
square feet of retail shops, dining, and entertainment space; and 1,650 luxury hotel, condominium, and private residency club units.

Atlantic City

When the civic leadership of Atlantic City first considered legalizing casino
gambling in the early 1970s, the city was in severe decline. Between 1960
and 1970, it had lost more than 20 percent of its population—dropping from
59,544 to 47,823. Unemployment and welfare rolls were growing, and older
neighborhoods were in decay with many abandoned buildings. Worse yet,

the city was losing its place as a major resort center, showing nearly a 40 percent decline in hotel capacity. Revenue from real estate and local luxury taxes was rapidly receding. Atlantic City was becoming a place primarily for the elderly and minorities.

With an eye on the success of Las Vegas, Atlantic City residents came to believe that casino gambling would be the magic bullet to reverse the city's declining fortunes. After failed attempts in 1970, 1973, and 1974, the voters of New Jersey in 1976 approved a referendum authorizing casino gambling in Atlantic City. The first casino opened on May 26, 1978. Five years later, in a study titled *The Atlantic City Gamble*, George Sternlieb and James W. Hughes could find few benefits for residents:

> The scorecard of winners and losers since Resorts International, the first of the Atlantic City's legal gambling palaces, opened its doors in 1978 can be stated very simply: the booming casino business in New Jersey is evident, but legalization of gambling has had negative results as well, among them increased activity by organized crime and a toll in human and economic displacement.[35]

From the very beginning, the returns in casino revenues and profits in Atlantic City exceeded even the most optimistic expectations. Between May 26, 1978, and the end of 1981, nine casinos began operating, representing an investment of well over $1 billion. Sternlieb and Hughes report that by the end of 1982, the city's casinos had approached the $3.7 billion mark in total gross revenue from gambling since its inception. The casinos also generated a substantial growth in new jobs, employing approximately thirty thousand workers. However, there has been little spillover growth in noncasino employment, and the number of unemployed in the city has evidenced no significant change. Most new jobs have gone to suburban residents.

Consequently, much bitterness remains among residents about the toll casino gambling has taken on individual homeowners, businesses, and neighborhoods. Writing in the *Philadelphia Inquirer*, Amy S. Rosenberg reports:

> For all the money in town, the casino era has created all-inclusive gaming halls and led to the closing of restaurants, nightclubs, movie theaters and stores, and thriving business districts turned into lackluster collections of barbershops, pawn shops, nail salons, delis and massage parlors. Local musicians who thrived in the old days have been all but shut out by casinos.[36]

Despite the bad rap, the city keeps trying to show that benefits have been trickling down to residents. Millions of dollars derived from a 1.25 percent investment tax on casino revenue have been spent on building and rehabilitating homes. Various neighborhoods boast new street signs and lamps, benches, trees, and sidewalks; and for some extra entertainment, new minor league baseball and basketball teams are now based in the city. The Casino Redevelopment Authority (CRDA), the state agency that administers the tax, has spent $880 million on community projects across New Jersey. For the greater region, the casino era has been an economic bonanza, with jobs going to people in counties and communities around Atlantic City.

In his book *Boardwalk of Dreams: Atlantic City and the Fate of Urban America*, Bryant Simon concludes that the proposed massive redistribution of wealth into the neighborhoods didn't happen. As he explains:

> That leaves Atlantic City pretty much as it is, a stark, vacant, poor city with a beach, the Boardwalk, and now that the Borgata [a new casino-hotel] has opened, twelve separate, inward-looking casino villages that leave only crumbs on the Monopoly streets around them.[37]

EDGE CITIES

The concept of "edge cities" is used here as a broad grab-bag category to account for the widespread growth of relatively high-order multifunctional centers that have emerged in the outer suburban areas of the nation. In his book *Edge City: Life on the New Frontier*, Joel Garreau explains that "Americans are creating the biggest change in a hundred years in how we build cities."[38] Where urban regions in the United States were once viewed as dominant central cities surrounded by constellations of suburbs, metropolitan areas are evolving as multiple centers made up of old central cities and other newer concentrations of edge cities. These have become self-sustaining economic entities in their own right. The old ecological model of urban regions where workers commute to the central city from dormitory suburbs is now out of date: today, only about a third of all suburbs are exclusively residential.

Typically located at the intersection of a beltway with a lateral road, edge cities surfaced in the 1950s as the result of advancements in automobile and communications technology. Garreau distinguishes three edge city types: *uptowns*, built on top of pre-automobile settlements that have been absorbed by urban sprawl; *boomers*, classic edge cities usually located at the intersection of freeways and almost always centered on a mall; and *greenfields*, representing state-of-the-art developments at the site of several thousand acres of farmland and one developer's monumental vision on a grand scale. Garreau identifies seventeen edge cities in Greater New York City and sixteen each in the Washington, D.C., and Los Angeles areas.

There are now more than two hundred such edge cities in various corners of the country.

Among the most impressive of these centers are Tysons Corner in Fairfax County, Virginia, in suburban Washington, D.C. (discussed later); the Galleria in suburban Dallas, which has three levels of shopping surrounding an ice rink, a luxury hotel, several restaurants and movie theaters, structured parking, and two major office towers with about one million square feet of space in one high-density setting; and Las Colinas, also in suburban Dallas, which serves as the world headquarters of Exxon (which moved from New York City in 1991). GTE Corporation has relocated to Las Colinas by way of consolidating its headquarters from Connecticut and other locations. This mixed-use complex includes housing, a people-mover system connecting commercial centers, and a series of lakes and canals offering water-based ferry service. To be noted also, in contrast to the low-rise building of most of these centers, the City Post Oak in suburban Houston houses the tallest office building in suburban America, the sixty-five-story Transco Tower.

An ongoing debate among those who favor edge cities and those who are critical of them concerns whether they are real communities as measured by social relations and established traditions. For the most part, the ways people relate to one another in these places is determined not by proximity but by cellular phones, e-mail, and fax machines. To the extent that residents can identify a common center, it is typically the shopping mall. After a century of evolution, malls now contain post offices, hotels, condominiums, counseling centers, and even amusement parks—all in safe, clean, weather-protected places.

Not everyone agrees, however, that malls are simply a new version of old central city downtowns. Kenneth Jackson notes in *Crabgrass Frontier* that because they are so self-contained and so uniform, shopping malls in fact represent almost the opposite of what downtowns really are: "They cater exclusively to middle-class tastes and contain no unsavory bars or pornography shops, no threatening-looking characters, no litter, no rain, no excessive heat or cold."[39] They are usually devoid of civic activity or anything that could be labeled serious culture such as theater, libraries, or concerts. As malls get bigger, however, like the Mall of America in Bloomington, Minnesota (see box 4.11), they are able to offer a greater array of customized products as well as special events.

Moreover, Garreau explains that edge cities do not have normal politics but are usually administered by what he calls "shadow governments." Organized like corporations, they may be privately owned and operated (e.g., homeowners associations) or they may be administered by quasi-public institutions (e.g., special districts or public-private partnerships). What makes them like governments is that they can assess mandatory

---

Box 4.11.   The Mall of America

The Mall of America in Bloomington, Minnesota, opened in 1992 as the nation's largest mall. However, the mall is not the largest in the world. Four larger shopping malls have been built in China and in North America. The West Edmonton Mall in Canada is larger, and Woodfield Mall in Schaumburg, Illinois, has more shopping space. The Mall of America is, however, the most visited shopping mall in the world, with more than forty million visitors annually. It has a total area of 4.2 million square feet, encompassing 520 stores, a seven-acre indoor amusement park that sports a roller coaster and a log ride, a wedding chapel (the Chapel of Love), an underwater adventure aquarium, a fourteen-screen movie theater, and a Lego imagination center, in addition to bars and nightclubs. In 2005, in Phase II, work began on the development of land north of the mall. This phase includes constructing a concert hall, a ski slope, an ice rink, condominiums, and a theme hotel.

---

fees, create rules and regulations, and coerce cooperation. However, their leaders are rarely accountable to all the citizenry in a general election, and they are usually not subject to the constraints the Constitution imposes on conventional governments.

Tysons Corner

The Tysons Corner story begins with the construction of a road built in the early 1960s with funds from the new National Defense Interstate Highway program. The road was intended to serve as a bypass around Washington, D.C., to better facilitate travel for people driving through to the western part of the nation. During the Cold War, the road was viewed as an asset to the nation's defense posture. Designed roughly as a circle sixty-six miles in circumference, it skirts the city by more than ten miles.

Local observers report that the land being built was so vacant that highway engineers attached no importance to the way the highway would cut across two dusty farm roads—Routes 7 and 123—just east of their intersection (which was named after a nineteenth-century landowner, William Tyson).

It didn't register on them that the resultant triangle would instantly become a place easily driven to and from any direction in the region. This superhighway was designed to get people from Maine to Florida. Local traffic? What local traffic? No commercial development was even imagined at any interchange.[40]

Nobody could foresee that the Beltway would become Washington's Main Street or that a major airport, Dulles International, would be built only a few miles to the west.

By 1960 the population of Fairfax County in Northern Virginia had tripled to more than a quarter of a million people in less than a decade. Residential subdivisions proliferated. The first rezoning for commercial purposes occurred in 1962 and continued at a rapid pace. Subsequently, in 1966 a federal grand jury indicted fifteen people under the Federal Racketeering Act on charges of conspiracy to exchange bribes for rezoning in Fairfax County. Persons who were targeted included county supervisors, developers, zoning attorneys, and one former state senator. While the senator was excused from trial for failing health, several others were sent to jail.

This episode was only a temporary setback to developing the area, however, because attempts by the County Board of Supervisors to manage growth in Fairfax were generally nullified in the courts on the basis of "property rights." By the 1980s, having surpassed Washington, D.C., in total population, Fairfax had become one of the larger and wealthier local jurisdictions in the country. According to Garreau, "Five formidable Edge Cities rose there," the most formidable being Tysons Corner—the largest urban agglomeration between Washington and Atlanta.[41]

Today, Tysons Corner boasts 31 million square feet of office space, ten hotels with 3,500 hotel rooms, and six million square feet of retail space, including the super-regional malls Tysons Corner Center and Tysons Galleria. Located at the eastern edge of the technology-rich Dulles corridor, it is home to one of the world's largest concentrations of innovative companies serving advanced government and corporate markets.

When a diplomatic contingent from China arrived in Washington in the early 1970s, its members put off a tour of such capital sites as the Lincoln Memorial and the Washington Monument and requested, instead, a visit to Tysons Corner to see "Bloomie's"—that is Bloomingdale's. This was the America that most piqued their curiosity.

SUMMARY AND CONCLUSIONS

In the new international division of labor, advancements in technology are creating new types of cities. Each city type offers distinct and unequal circumstances for residents in terms of earnings, services, and overall quality of life. Third World border cities that are linked primarily to Latin American countries typically have large numbers of low-paid workers. Many such places are cities where routine manufacturing takes place on the foreign side of the border and headquarters and high-tech activities are situated on the American side. Old industrial centers also function as places for routine economic tasks and thus are increasingly dependent on control centers located in other parts of the nation and the globe. Many are old factory cities of the Northeast and Midwest that are seeking new roles in the changing economic order. In the meantime, most such places experience fiscal duress. Similarly, many retirement communities depend on outside sources of income in the form of pension, society security payment, and federal and state programs to support the local economy.

Leisure-tourism playgrounds tend to have a two-tier economy consisting of many low-wage service personnel and comparatively few high earners who make their money as real estate speculators or top managers. For more and more such places, urban tourism in the form of glitzy gambling casinos and/or high-tech theme parks have come to play a key role in promoting economic development, in some instances—as in the case of Atlantic City—with mixed results.

Innovation centers perform research and development functions, and their products are then carried into other worldwide settings. Heavy reliance is placed on continuous inputs of knowledge for advancing computer technology. Here, highly skilled workers and entrepreneurs enjoy upscale living standards and many become millionaires. On the downside, real estate and rental values typically price low-paid service workers out of the market. In comparison, headquarters cities attract top corporate decision makers from finance, industry, commerce, law, and media. At the highest level, cities such as New York, Chicago, and Los Angeles function as world command centers where multinational corporations create a truly transnational economy as facilitated by communications technology. But such cities also evidence a dual economy—the very rich, who have entrepreneurial and technical skills, and the very poor, who lack such skills and find it difficult to compete in the job market.

Meanwhile, more and more Americans continue to move to the suburbs, where land is abundant and relatively cheap and where the tax code allows households to deduct mortgage interest property taxes for both first and second homes. In the 1990s, the United States became the first nation in history to have as many suburbanites as city and rural dwellers combined. Where suburban communities formerly served as bedroom communities to the central city, they are now being transformed to mixed-use multifunctional centers that act as edge cities—here malls serve as the new versions of a downtown. As development has continued to ooze into the hinterlands of metropolitan regions, urban sprawl and the loss of open space have become hot issues almost everywhere.

NOTES

1. Robert B. Reich, *The Work of Nations: Preparing Ourselves for Twenty-First-Century Capitalism* (New York: Vintage Books, 1992), 3, 8.

2. Paul I. Knox, "Globalization and Urban Economic Change," in *Globalization and the Changing U.S. City*, ed. David Wilson (Thousand Oaks, Calif.: Sage, 1991), 21.

3. John R. Logan and Harvey L. Molotch, *Urban Fortunes: The Political Economy of Place* (Berkeley: University of California Press, 1987), 258.

4. Saskia Sassen, *Cities in a World Economy*, 3rd ed. (Thousand Oaks, Calif.: Pine Forge Press, 2006).

5. Peter J. Taylor and Robert E. Lang, "U.S. Cities in the 'World City Network'" (Washington, D.C.: Brookings Institution, 2005), available at www.brookings.edu/metro/pubs/20050222_worldcities.pdf.

6. Taylor and Lang, "U.S. Cities," 3.

7. See Janet L. Abu-Lughod, *New York, Chicago, Los Angeles: America's Global Cities* (Minneapolis: University of Minnesota Press, 1999).

8. J. Kaplan, "Rooting for a Logo," in *The City and the World: New York's Global Future*, ed. Margaret E. Crahan and Alberto Vourvoulais-Bush (New York: Council on Foreign Relations, 1997), 159–70.

9. U.S. Bureau of Labor Statistics, "9/11 and the New York City Economy," *Monthly Labor Review* 127 (June 2004): 1, 2.

10. John Hull Mollenkopf and Manuel Castells, eds., *Dual City: Restructuring New York* (New York: Russell Sage Foundation, 1991), 402.

11. Allen J. Scott, *Technopolis: High-Technology Industry and Regional Development in Southern California* (Berkeley: University of California Press, 1993).

12. Michael Dear and Steven Flusty, "The Iron Lotus: Los Angeles and Postmodern Urbanism," in *Globalization and the Changing U.S. City*, ed. David Wilson (Thousand Oaks, Calif.: Sage, 1997), 163.

13. Stuart A. Gabriel, "Remaking the Los Angeles Economy," in *Rethinking Los Angeles*, ed. Michael Dear, H. Eric Schockman, and Greg Hise (Thousand Oaks, Calif.: Sage, 1996), 22–33; Brookings Institution, Center for Urban and Metropolitan Policy, *Los Angeles in Focus: A Profile from Census 2000* (Washington, D.C.: Brookings Institution, 2003), available at www.brookings.edu/es/urban/livingcities/LosAngeles.htm.

14. Gabriel, "Remaking the Los Angeles Economy," 26.

15. Edward Soja, "Urban Restructuring in New York and Los Angeles," in Mollenkopf and Castells, *Dual City*, 372–73.

16. Soja, "Urban Restructuring," 373.

17. Edward W. Soja, *Postmodern Geographies: The Reassertion of Space in Critical Social Theory* (London: Verso, 1989), 192.

18. Michael Lewis, "The Little Creepy Crawlers Who Will Eat You in the Night," *New York Times Magazine*, 1 March 1998, 43.

19. Martin Kenney, ed., *Understanding Silicon Valley: The Anatomy of an Entrepreneurial Region* (Stanford, Calif.: Stanford University Press, 2000).

20. Lisa M. Montiel, Richard P. Nathan, and David J. Wright, *An Update on Urban Hardship* (Albany, N.Y.: Nelson A. Rockefeller Institute of Government, 2004).

21. Rod Gurwitt, "Black Mayors and the Board Room," *Governing* 7 (April 1996): 42.

22. Logan and Molotch, *Urban Fortunes*, 272.

23. Logan and Molotch, *Urban Fortunes*, 273–75.

24. Logan and Molotch, *Urban Fortunes*, 275.

25. Brookings Institution, Center for Urban and Metropolitan Policy, *Growing the Middle Class: Connecting All Miami-Dade County Residents to Economic Opportunity* (Washington, D.C.: Brookings Institution, 2004), available at www.brookings .edu/metro/publications/20040607_miami.htm.

26. Charles F. Longino Jr., *Retirement Migration in America*, ed. R. Alan Fox (Houston: Vacation Publications, 1995).

27. Susan A. MacManus, with Patricia A. Turner, *Young v. Old: Generational Combat in the 21st Century* (Boulder, Colo.: Westview Press, 1996).

28. Maria D. Vesperi, *City of Green Benches: Growing Old in a New Downtown* (Ithaca, N.Y.: Cornell University Press, 1985), 39.

29. "Tourism Industry Profile," National Assembly of State Arts Agencies, www.nasaa-arts.org/artworks/profile.shtml.

30. Ada Louise Huxtable, *The Unreal America: Architecture and Illusion* (New York: New Press, 1997).

31. Ada Louis Huxtable, "Living with the Fake and Learning to Like It," *New York Times*, 30 March 1997.

32. John M. Findlay, *Magic Lands: Western Cityscapes and American Culture after 1940* (Berkeley: University of California Press, 1992).

33. Huxtable, *Unreal America*, 48.

34. David Schwartz, *Suburban Xanadu: The Casino Resort on the Las Vegas Strip and Beyond* (New York: Routledge, 2003).

35. George Sternlieb and James W. Hughes, *The Atlantic City Gamble* (Cambridge, Mass.: Harvard University Press, 1983), 10.

36. Amy S. Rosenberg, "Casinos' Blessings Have Been Mixed," *Philadelphia Inquirer*, 23 May 2003.

37. Bryant Simon, *Boardwalk of Dreams: Atlantic City and the Fate of Urban America* (New York: Oxford University Press, 2004), 222.

38. Joel Garreau, *Edge City: Life on the New Frontier* (New York: Doubleday, 1991), 3.

39. Kenneth T. Jackson, *Crabgrass Frontier: The Suburbanization of the United States* (New York: Oxford University Press, 1985), 23.

40. Garreau, *Edge City*, 377.

41. Garreau, *Edge City*, 385

5

Conclusions:
Cities in the Third Wave

By the early 1900s, the industrial city had evolved into a highly complex urban settlement of diversified industrial-commercial economies. Comprehending this phenomenon became a major challenge confronting urban theorists of the time. To meet this challenge, University of Chicago sociologists Robert Park, Ernest Burgess, and Roderick McKenzie introduced the concept of *urban ecology*.[1] In explaining how communities evolve, they stressed the role of competition, comparing it to that which plays out in the struggle for survival in the world of nature. Competition in the industrial city was a struggle between competing groups for physical and social access to desirable spatial locations in which to implement diverse economic, residential, and social activities. Metaphors of invasion and succession were invoked to represent the filtering out of populations from older locations around the central business district to newer outlying areas.

In most large cities, the central business district was the dominant core of the metropolis, containing corporate headquarters, major banks, retail stores, and other business, commercial, and entertainment entities. Surrounding this district was a series of concentric zones of different land uses, each tending to encroach on others. Amos Hawley describes the process as follows:

> Growth of the central business district pushes ahead of a belt of obsolescence occupied by light industries, warehouses, and slums. This transition zone, in turn, encroaches on a zone of low-income housing, causing the latter to shift outward and to invade a belt of middle-income residential properties. . . . The

occupants of each inner zone tend to succeed to the space occupied by those of the next outer zone. At any moment in time, therefore, the distribution of land use exhibits a ring-like appearance.[2]

In certain situations topographical features such as hills, rivers, and lakes were viewed as influencing these zones, creating odd patterns, which Home Hoyt labeled "sectors."[3] Transportation systems—for example, railroads, highways, and boulevards—also interrupted the zonal pattern.

After the 1930s, urban ecologists tended to refine their early theoretical perspectives on classic ecology and narrowed the focus to the spatial distribution in urban areas of different behavioral problems such as crime, delinquency, and mental disorders.[4] In addition, demographic studies were made of different types of population movements, such as migration from rural to urban areas and migration from cities to suburban areas after World War II. As described by Brian Berry and John Kasarda, "A majority [of urban ecologists] shifted their attention to the narrow and somewhat unimaginative graphing and mapping of social data to uncover spatial correlations."[5]

At the present time, urban theorists are again confronted with the challenge of understanding how cities develop. As was discussed in the previous chapters, since the 1950s metropolitan areas have been subject to an unprecedented degree of restructuring. In accounting for such change, we have tried to show how technological innovation is creating new kinds of urban conglomerations. Thus, contemporary inquiry now extends beyond the expansion of the central city into the outlying suburban zone as an ecological phenomenon and focuses instead on major interregional and international shifts of facilities, jobs, and investment. As a result, postmodern urban theory is a theory of the reorganization of the urban space economy (as driven by post-postindustrial forces of change) at the regional, national, and international levels. To focus on how cities adapt to such change, a whole new agenda of questions must be examined, including the role of growth coalitions in managing change, the role of "citistates" in the global economy, cities in pursuit of niche markets, and cities as entertainment centers.

On a broader level of interest, there are issues of livability and sustainability. Given the changes that are taking place today, and particularly the way people use communications technology, some theorists claim that urban agglomerations are too large and too fragmented to support viable communal relationships. At the same time, urban spread is viewed as being environmentally damaging. Thus, a basic concern addressed here is whether postmodern urban society is sustainable. This includes the question of what will become of core cities and older metropolitan areas.

THE ROLE OF GROWTH COALITIONS IN CITIES

One key implication, according to Mark Gottdiener and Joe Feagin, is the need to give greater attention to group and individual actions within urban contexts.[6] That is to say, in addition to recognizing structural factors dictated by technology and a free market, it is important to understand the role of agency by observing how powerful actors operate to manage urban growth. John Logan and Harvey Molotch contend that "the activism of entrepreneurs is, and always has been, a critical force in shaping the urban system, including the rise and fall of given places."[7] Some analysts use the term *urban regime*, recognizing that local government alone cannot mobilize and coordinate the necessary resources to govern and compete.[8] Thus, business leaders typically play significant roles in the city's informal governing coalition: "Government wants business investment, but business also wants government cooperation."[9] In addition to politicians and business leaders, key players in local growth coalitions include local newspapers, leaders of utilities, and the transportation bureaucrats, both public and private.

In this light, urban restructuring is viewed as a combination of changing economic conditions driven for the most part by technology and the capacity of urban elites to deal effectively with capital. Furthermore, the issue of who dominates is understood to produce an uneven capacity to attract growth, which in turn renders different states of advantage and disadvantage among different groupings of people.

To illustrate, San Francisco is a city that has had to reinvent itself not once but many times. In the 1920s, San Francisco was a brawling port city, the premier corporate center of the western United States. By the 1930s, however, the city's status was challenged by Southern California's rapid population growth. Unable to compete with Greater Los Angeles in size and economic energy, San Francisco's business leaders in the 1950s and 1960s began to emphasize the city's lifestyle. Looking to create a highly compact alternative to New York and other traditional financial centers, the city's corporate elite envisioned an upscale metropolis boasting of its own transit-centered high-rise complex, luxury apartments, and fashionable shopping.

By the 1970s Manhattanization, as the critics labeled it, clashed with the determined opposition of counterculture advocates and neighborhood activists. In the late 1980s, antigrowth forces succeeded in imposing strict limits on high-rise development in the city. The antigrowth movement preserved the city's physical charms, at the cost of its business appeal. But certain countertrends set in as well. Under a succession of mayors and a Board of Supervisors disdainful of economic development, the city seemed unable to manage its growing homeless population, keep its streets clean, or run its municipal transit system. Between 1990 and 1993, nearly forty thousand jobs were lost.

With the election of Willie Brown as mayor in 1997, a new growth coalition came to control city politics. However, rather than attempting to re-create Manhattan, the city decided to pursue a twofold strategy—promoting tourism, the city's largest industry, and supporting the development of financial and legal services and multimedia and software businesses tied to the growth of high technology in neighboring Silicon Valley. In essence, the city has been marketing itself as a preferred lifestyle for postindustrial elites. But in taking this route, San Francisco appears to be abandoning its blue-collar heritage and much of its economic diversity. Its once busy port has been supplanted by Oakland's; export business has shifted toward San Jose, which exports four times as much as San Francisco. Other blue-collar industries have lost ground, and scores of smaller manufacturing have departed.

"CITISTATES" IN THE GLOBAL ECONOMY

An important feature of postindustrial urbanization, as we have noted, is the emergence of metropolitan regions as critical participants in the world economy. Neal R. Peirce has chosen the word *citistates* to emphasize these entities' transglobal connectedness and their growing domination on the national and international levels.[10] Several factors can account for this:

- Advancements in telecommunications now allow investment in capital and other resources to move rapidly and freely anywhere in the world, with the capacity to aggregate and disaggregate in key locations.
- Trade agreements such as the General Agreement on Tariffs and Trade (GATT) and the North American Free Trade Agreement (NAFTA) have weakened trade barriers and liberalized the flow of goods across national boundaries.
- At the microeconomic level, metropolitan regions now serve as sources of entrepreneurial leadership, even to the point of eclipsing the nation-state.

According to Saskia Sassen, the transformation of the world economy during the last two decades and the accompanying shift to services and finance has renewed the importance of major cities.[11] The combination of global banking, financial services, and corporate headquarters in and around the core cities of key metropolitan regions establishes them as high-level command and control centers over an ever-widening hinterland.

Various urban theorists contend that the rapid integration of the global economy will be among the most important factors shaping urban economics in the United States.[12] In order to compete in the national and international economy, cities and regions must demonstrate the availability of certain key assets, including a better educated and more highly skilled workforce; globally linked telecommunications; efficient air and surface transportation; knowledge-based research institutions; flexible, mission-oriented public and private organizations; an attractive quality of life; and fiscal soundness.[13]

To illustrate: Austin, Texas, has been able to attract international companies largely because of its skilled and highly educated workforce. Similarly, Salt Lake City provides an attractive labor market because it has the highest literacy rate in the United States. Columbus, Ohio, and Baltimore have attracted new medical and technical firms because their universities are viewed as critical assets in the production process. The Raleigh-Durham region qualifies as a blossoming competitor not only on the basis of the quality of its workers and technical talent but also as one of the best places in the United States to live and work. The Minneapolis–St. Paul region has lured companies because of its reputation for a high quality of life—it is among the nation's cleanest and safest metropolitan areas, known for its rich cultural offerings.

To maintain economic competitiveness, other cities have focused on developing their telecommunications infrastructures. As more economic functions are conducted electronically, the ability to transmit and receive large amounts of information rapidly is a critical factor in national and international competition. During the 1980s, a growing number of municipalities viewed teleports as akin to airports and maritime ports as spurs to economic development. A teleport is basically a user or provider that consolidates communications to and from satellites. For example, the Washington International Teleport is an intermediary that moves mostly video information from earth to satellite and back. Because of its location on the Mid-Atlantic coast, it is an information gateway for countries on the Atlantic rim. Another example is Kalamazoo, Michigan, which designed a Telecity USA project. Telecity joined a group of local communications networks linking local businesses, institutions, and citizens to the National Information Infrastructure (NII), allowing participants to share information with approximately 85,000 households. Participating businesses include Upjohn Corporation and First America Banking Corporation.

Especially important in maintaining economic competitiveness is the extent to which the public, private, and civic sectors cooperate in adapting to change. The reorganization of Pittsburgh's regional economy, for example, can be attributed mainly to the close cooperation among business executives, state and local government officials, and university officials. A metropolitan system of governance or planning, furthermore, allows metropolitan areas to project a credible image to outside firms by sending the message that these areas can draw on a wide range of knowledge, resources, and perspectives.

To date, the most ambitious effort to achieve regional change in North America is Toronto's federated metropolitan government. In 1953 Toronto established a "two-tier" federation, which consists of a single, areawide government (the first tier) and preexisting local governments (the second tier). The first tier has jurisdiction over the entire metro area in such important urban functions as water supply, sewage disposal, mass transit, health services, arterial roads, public housing, and planning. The second tier focuses on police, fire control, sanitation, and education. Short of federation, another model of structural change is the comprehensive urban county plan. A two-tier government operating in (Miami-) Dade County, Florida, since 1957 gives the county government a powerful and integrating role over an area comprising 2,054 square miles and twenty-seven municipalities. Among its responsibilities are mass transit, expressways, health and welfare services, and communication for fire and police protection.

CITIES IN PURSUIT OF NICHE MARKETS

Alongside the new citistates are cities that lack some or all of the key features that would allow them to participate more fully in the new global economy. Such cities tend to become peripheral, increasingly excluded from the major economic processes that fuel growth. Old industrial cities and border cities appear to be most vulnerable. In the struggle to maintain a viable economic base, such cities strive to restructure their economies via the creation of niche markets such as tourism, gambling, convention centers, back-office operations, and other options that could draw new investment. They typically depend on the presence of low-wage labor and cheap land to promote economic development.

Most major cities across the nation have seen a boom in the building of convention centers, usually accompanied by downtown hotels and shopping centers. Growth in tourist-oriented construction in these places has actually outpaced the growth of commercial office space. In 1970, convention centers in the United States offered some 6.7 million square feet of exhibition space, and only fifteen cities could handle a trade show for twenty thousand people or more. Twenty years later, in 1990, convention centers covered more than 18 million square feet of exhibition space and 150 cities were able to accommodate trade shows for twenty thousand people.[14]

In almost all cases, the construction of convention centers has been funded by the public. Under the federal urban renewal program of the 1960s, cities used federal aid to clear downtown sites for new convention centers. Subsequently, under the Urban Development Action Grant program, cities built hotels and convention centers in an effort to lure conventioneers and other visitors to their downtowns. With the decline in federal aid, cities are now expected to come up with the necessary dollars by floating bonds or imposing nuisance taxes. For example, the $987 million expansion of McCormick Place in Chicago was financed through general obligation bonds, and debt service was paid with proceeds from the state cigarette tax. New York's convention center was financed through a $375 million bond issue.

Studies show that, while convention centers themselves rarely make a profit on the conventions and trade shows they host, the dollars conventioneers typically pay for food, lodging, transportation, and incidentals help prop up the city's economy. However, there remains the question of how much cities actually benefit from such development. In his assessment of new expanded centers in five cities—Los Angeles; Washington, D.C.; Providence, Rhode Island; Philadelphia; and Boston—Heywood T. Sanders reports that they fall short of what the feasibility studies promised. One major constraint is competition from other cities building ever larger centers to steal away convention goers and visitors. Sanders elaborates:

> For Atlanta, with one of the largest centers in the nation, expansions in Chicago, New Orleans, Washington, and Los Angeles threaten to leave the city with only 7 percent of the total exhibition space among the Major U.S. Centers by the year 2006 if it does not expand. Washington, D.C., now lags behind the centers of the cities with which the District principally competes, having fallen from fifteenth to twenty-fifth nationally in the total amount of space offered. The competition is no less serious for San Antonio, where virtually all of its primary convention-business competitors have either recently expanded, are in the process of expanding, or have plans to expand within the next few years.[15]

Adding casino gambling to attract tourists is another increasingly prominent marketing strategy. East Saint Louis, Illinois, which lies just across the Mississippi River from Saint Louis, shows that a dying city can seize on riverboat gambling to survive and make a comeback. This city of 31,000 people is predominantly black, and 75 percent of its population lives on some form of welfare. The U.S. Department of Housing and Urban Development described it as "the most distressed small city in America." In the 1930s, industries that had enticed black people to come to East Saint Louis with promises of jobs began to move to places where even cheaper labor could be found. Proximity to coal, which had attracted industry to the area, ceased to be important as electric power became commercially available in other regions. The city underwent a resurgence in World War II when available factory space was used for military manufacturing. Population peaked in 1945 at 80,000—one-third black. By 1971, with the population down to 50,000—less than one-third white—a black mayor was elected. That has been the pattern ever since.

Before riverboat gambling was introduced in East Saint Louis in 1993, raw sewage was backed up into the streets and mounds of trash were piled near vacant and burned-out buildings. When there was not enough money to buy gas for police cars, the crime rate escalated and murder was rampant. Since the side-wheeler *Casino Queen* began cruising in the shadow of the Gateway Arch, it has generated more than 1,200 jobs. The city has used the approximately $9 million it receives in annual casino tax revenue to begin reversing its long decline. Broken sewers have been repaired, streets have been paved, and more than twelve hundred derelict buildings have been demolished. The city's police force has nearly doubled, and its firefighters have received more than $1 million in new equipment. Because the crime rate has gone down, it is estimated that the nearly three million people who visit the casino each year actually venture into the city where—until recently—people feared for their lives.

Other towns and cities clamor for the boom that a big prison brings. Residents of most communities shudder at the thought of having a prison built in their neighborhoods. But that feeling is not shared in many economically depressed towns of rural New York where a new prison is as welcome as a new factory, department store, or office. People in places like Johnstown, west of Albany, and Romulus, a community sixty miles southeast of Rochester, saw new opportunities when Governor George Pataki proposed building three maximum-security state prisons at a cost of $476 million.

At the lowest level, niche markets serve as dumping grounds for refuse produced elsewhere. To generate revenue, poor communities in Pennsylvania and Virginia provide landfills for out-of-state trash from New York and other Northeastern cities. In these two states, solid-waste enterprise generates millions of dollars a year in fees for the state and local communities. However, as illustrated in box 5.1, social costs are usually borne by those at the bottom of the class ladder.

Some areas show more ingenuity in managing landfills than others. For example, the Hackensack Meadows, which for many years served as landfills for New York City's garbage, found a new way to make money. In 1971, the Hackensack Meadowlands Development Commission went to court to prevent further expansion of landfills, and it won the case in 1978. The remaining landfill sites, in Kearny and North Arlington, New

---

Box 5.1. Laid to Waste

In the economically depressed, largely African-American "West End" of Chester, Pennsylvania, Zulene Mayfield lives next door to the fourth-largest trash-to-steam incinerator in the nation and a few doors away from a large processing facility for infectious and hazardous medical waste. The county's sewage treatment plant sits adjacent to her neighbor's homes a block away, and additional toxic-waste processing facilities have been proposed for the community.

Every day, trucks from as far away as Virginia roll past homes on Chester's Second Street, delivering thousands of tons of waste. Residents believe that their lives are being disrupted, their health threatened, their community destroyed, and the very air they breathe dangerously polluted. A grassroots organization called Chester Residents Concerned for Quality Living (CRCQL) has taken an active role in opposing the facilities and in publicizing the plants' impact on their community. Representatives of the waste-processing companies argue that their facilities are safe and that they bring much needed jobs to Chester.

Source: *Laid to Waste*, video (Berkeley: University of California Extension Center for Media and Independent Living, 1997).

Jersey, were about 150 feet high before the dumping was halted. The hills made for conditions suitable for the recovery of methane gas, which is created as garbage decomposes. The Garbage Alps, as they are called, produce six million cubic feet of methane gas a day, enough to meet the energy needs of as many as ten thousand homes. The commission contracted with Air Products and Chemicals, Inc., which invested in a $12 million compression plant to bring up about $5 million worth of gas annually. In return, the commission receives a 10 percent royalty on the gas, and that royalty covers about 10 percent of its operating budget.

CITIES AS ENTERTAINMENT CENTERS

Another growth strategy that has become commonplace in urban areas features high-tech fun. As we have noted, a staggering array of ambitious projects are going up across the country—sports arenas, cultural centers, entertainment-enhanced malls, urban theme parks, and entertainment-enhanced retailing. The expectation is that entertainment zones can help restore some of the luster to America's central cities, bringing back some of the economic and social magnetism they have been losing to the suburbs. From Cleveland's Rock and Roll Hall of Fame to Universal Studio's City-

Walk mall in Los Angeles to the overhaul of New York's Times Square, the point is being demonstrated that a central location can still be advantageous for entertainment. In contrast to the isolating effects of home video and the Internet, urban entertainment zones provide social interaction.

The trend toward converting cities into places of entertainment as a means of generating revenue stands in contrast to city life in the late nineteenth and early twentieth centuries. At that time, the most ambitious undertakings in America were primarily public civic centers, parks, and major cultural institutions (e.g., the San Francisco Civic Center and Opera House, the Art Institute in Chicago, and the New York Public Library and Central Park in Manhattan). Today, large-scale projects are, for the most part, privately sponsored and publicly subsidized, and the greater share of profits are returned to the investors.

Professional sports offer one illustration. In earlier times, most professional teams played in privately built stadiums, but this began to change in the 1950s. Cities built stadiums with public monies, and this soon became a negotiating point for any team that showed an inclination to move to another city. To be noted is that teams have leverage in negotiating with cities because professional sports are cartels, and cartels can limit the number of available teams to artificially boost their value. This applies to most professional sports. Today a team can usually keep all the money it takes in from luxury boxes, stadium advertising, parking, and concessions. Although each sport has its own rules, TV broadcasting revenues and gate receipts are generally shared with the rest of the league.[16] As an example, table 5.1 provides financial data for new baseball stadiums.

IS THIRD WAVE URBAN DEVELOPMENT SUSTAINABLE?

As typically envisioned by urban planners, the ideal city is a place that is compact and tightly integrated in its parts. Such a form of development, they contend, allows greater efficiency in its upkeep and, perhaps more important, conveys a sense of community whereby neighbor engages neighbor.[17] But as we've seen, Third Wave cities are far-reaching in scope and consume huge quantities of resources. In the search for a path for a sustainable society, it can be argued that they pose threats to the environment in addition to causing social and economic stress.[18] Consider the following:

- Massive and inefficient energy consumption that results from urban sprawl wastes resources and generates greenhouse gases.
- Motor vehicles dominate urban transportation systems, producing gridlock and pollution.

Table 5.1. New Ballparks and How They Were Financed (in millions)

Stadium	Team	Year Opened	Total Cost	Public Cost	Public Share
Oriole Park	Baltimore Orioles	1992	$314	$294	94%
Jacobs Field	Cleveland Indians	1994	$221	$221	100
Ameriquest Field	Texas Rangers	1994	$242	$204	84
Coors Field	Colorado Rockies	1995	$264	$207	78
Turner Field	Atlanta Braves	1997	$274	—	0
Bank One Ballpark	Arizona Diamondbacks	1998	$407	$291	72
Safeco Field	Seattle Mariners	1999	$601	$418	70
SBC Park	San Francisco Giants	2000	$359	$11	3
Minute Maid Park	Houston Astros	2000	$270	$184	68
Comerica Park	Detroit Tigers	2000	$327	$125	38
PNC Park	Pittsburgh Pirates	2001	$277	$227	82
Miller Park	Milwaukee Brewers	2001	$414	$324	78
Great American Park	Cincinnati Reds	2003	$353	$268	76
Citizens Bank Park	Philadelphia Phillies	2004	$460	$228	50
Petco Park	San Diego Padres	2004	$474	$301	64
Cardinals Stadium	St. Louis Cardinals	2006	$346	$208	60

Source: *Sports Business Journal*; Wisconsin Legislative Audit Bureau.

- The economies of cities reflect growing inequality of wealth and income distribution wherein some cities are winners and others are losers.
- Urban production and consumption extract resources from around the planet and deposit massive amounts of waste products.
- The emerging information society leads to separation among those who are able to master technology and those that lag (the dual city metaphor).
- The emerging informational society affects the way people relate to each other, with a growing tendency toward social isolation.

In light of these factors, the following discussion takes a more in-depth look at three key areas that bear upon the critical question of urban sustainability: urban spread as it affects the urban environment; people's sense of community as affected by information technology; and the changing role of core cities and older metros. Viewing the trends, what are the opportunities and what are the obstacles?

The Prognosis for Urban Spread and the Urban Environment

Over the past thirty years or so, many outer suburbs have evolved as the healthiest part of the metropolitan economy and the strongest part of the national economy. The prognosis for the immediate future is that the economy of these outer areas will continue to expand as jobs and firms continue to relocate out of the central city and inner suburbs.

An important feature of suburban spread, as previously noted, is the creation of edge cities and industrial parks. Edge cities are oriented to new, mixed-use concentrations of retail, office, and high-tech industries. Industrial parks are places that benefit from cheap land where factories and other production centers can be designed and built according to specifications. Lawrence Herson and John Bolland elaborate:

> Unlike the slow step-by-step growth of the older, traditional cities, both [edge cities and industrial parks] are very much the consequence of what has come to called "land developer's drive." In fairly quick sequence, site selection, financing, land acquisition, and then bulldozers and construction crews change the suburban landscape dramatically.[19]

To many advocates of smart growth, edge cities represented an alternative to ongoing sprawl. They anticipated that edge cities would progressively grow denser and more pedestrian friendly and eventually become established suburban city centers somewhat comparable to the downtown cores of older central cities. But this vision is increasingly overshadowed by a relatively new form of development: spatial leapfrogging as facilitated by highways and computer technology.[20]

In his book *Edgeless Cities: Exploring the Elusive Metropolis*, Robert E. Lang documents a decentralized form of office development growing beyond suburban boundaries.[21] As he observes, "edgeless cities" are made up of free-standing buildings, office parks, or small clusters of buildings of varying densities, strung along suburban interstates and major arterials. As of 1999, edgeless cities in the thirteen largest metropolitan office markets studied by Lang accounted for more than a third of all office space (36.5 percent). In some contrast, edge cities accounted for only 20 percent.[22] Lang observes that edgeless cities represent the newest phase in the continuing decentralization of the urban landscape.

Though not the only factor that has contributed to such spread-out development, motor vehicles continue to have a voracious appetite for land. Between 1970 and 1990, motor vehicle use in the United States doubled from one to two trillion miles. Consequently, in the 1990s, land consumption in the United States proceeded at twice the rate of population growth. As reported by the U.S. Department of Housing and Urban Development, land development between 1994 and 1997 averaged 2.3 million acres annually. Of the more than 9 million acres developed during those years, the overwhelming majority was outside metropolitan areas.[23]

Unfortunately, where land development is not carefully planned, the environment suffers. This can be illustrated in the case of Los Angeles. No other American city exhibits an environment in which the highest quality of urban life can be achieved at the same time that it is threatened with the greatest number of physical hazards. What makes Los Angeles so attractive is its semidesert climate of rainless summers and mild winters combined with enticing beaches and tropical vegetation. Snowcapped mountain peaks serve as backdrop. Yet the area is vulnerable to catastrophe because of the very same features that have attracted so many inhabitants. As more and more people settle in the area seeking a leisure-oriented good life, they have inadvertently exacerbated the potential for disaster through thoughtless development and indiscriminate use of natural resources.

In search of scenic beauty and appreciating land values in "unique" settings, Angelenos have built on top of cliffs, near ravines, along the coastline, in arid brushland, and next to geological faults and are thus vulnerable to brushfires, mudslides, drought, smog, flooding, and, most seriously, earthquakes. In fact the whole of Los Angeles is underlain by a fault mosaic that periodically causes tremors in the earth. In 1971, severe tremors brought down several overpasses in the city's freeway system and demolished or damaged more than one thousand buildings. Despite gloomy predictions by scientists of severe earthquakes in the future, the city's inhabitants tend to be fatalistic.[24]

Another example of uncontrolled development pertains to the explosive growth of population in Las Vegas, which long ago outstripped its own natural-resource infrastructure. In a desert basin that receives only four inches of annual rainfall, irrigation of lawns and golf courses only accelerates environmental deterioration. What water Las Vegas cannot buy from Arizona farmers, it seems determined to divert from the Virgin River. Furthermore, heavy use of automobiles in the desert environment has contributed to unhealthy air quality. Like Phoenix and Los Angeles before it, Las Vegas was once a mecca for people seeking the curative powers of clean air. Now, according to Environmental Protection Agency reports, Las Vegas is in a tie with New York City for fifth place in carbon monoxide pollution. Its smog contributes to haze over the Grand Canyon and is beginning to reduce visibility in California's new East Mojave National Recreation Area as well. Meanwhile, dune buggies, dirt bikes, speed boats, and jet-skis continue to tear away at the neighboring fragile desert environment.

Managing Growth

Though the two cases cited here represent extreme examples of environmental contamination, other U.S. communities are similarly unresponsive to the negative impact of development. In this light, an important question to be posed is how urban growth can be best managed to assure a sustainable environment.

Perhaps the best-known example is Portland's Smart Growth strategy. In 1973, the Oregon legislature passed an urban growth boundary law that requires each municipality in the state to draw a line beyond which urbanization could not go, except under very restrictive circumstances. Each of Oregon's 241 cities has come to be surrounded by an urban-growth boundary. Portland defined its boundary in 1979. Consequently, Greater Portland has not only stayed aesthetically pleasing but has also met the Smart Growth goal of increasing density.

However, while sprawl has been contained, it has not been stopped. The discouraging news is that the Portland urbanized area has continued to spread into rural areas. As people have continued to pour into Oregon, development pressures within the "containment vessel" of the urban-growth boundaries have been intensifying. Increasing numbers of residents have been decrying the added congestion and surging housing prices that are the result of trying to prevent sprawl in a time of rapid population growth.

Another interesting example is Maryland's Smart Growth policy. In 1997, then governor Parris Glendening unveiled a new statewide growth control policy that was hailed as a promising new tool for managing growth. The basic approach was simple: Public funds for infrastructure (e.g., roads, water lines, sewer lines) would not be used to support development in undeveloped areas of the state. Under the law, counties were to submit plans to the state showing where they intend to support growth. Designated "priority funding areas," these zones would be eligible for state infrastructure financial assistance, while projects that fell outside the boundaries of these areas would not be eligible for support.

Similar to the Portland case, the Maryland strategy was hailed as a national model for controlling growth. But what looked good in concept proved otherwise in practice. As reported in the *Washington Post*, there have been no significant shifts in Maryland's development patterns since the passage of the state's Smart Growth package.[25] According to Maryland's own planners, the rate at which farm and forest land has been destroyed has not slowed. One reason for the failure of the Smart Growth policy was that, although the state could refuse to fund the necessary public infrastructure, it could not veto a project. Large developers and retail giants such as Wal-Mart could finance the necessary roads and sewers themselves. As long as developers were paying their own way and the projects were viewed as benefiting the local economy, local officials were quite willing to give their endorsement.

Given the strong tradition of property rights in U.S. society and the very considerable influence that growth coalitions are able to assert, it is likely that any progress in controlling growth will be incremental at best. For example, former New Jersey governor Christie Todd Whitman successfully pressed for $1 billion in loans to set aside half of the state's remaining two million acres of open space over a period of ten years. Voters overwhelmingly approved the measure in November 1998. New York City's Land Acquisition Program (LAP) is another example of managing land development. In order to protect its water source in the Catskill-Delaware watershed, the city has committed to soliciting a minimum of 355,000 acres of land over a ten-year period. The basic approach is for the city to purchase fee title or acquire conservation easements on property that is considered to be water-quality sensitive. Of some significance also are the activities of the Nature Conservancy, a nonprofit organization that has been buying environmentally sensitive land areas both within the United States and worldwide for purposes of preservation.

The Prognosis for a Regional Approach to Urban Development

The challenge of dealing with environmental externalities (i.e., spillover effects such as air or water pollution) on the local level continues to be one of the most formidable obstacles to attaining sustainable cities. Most environmental problems are too big and expensive to be tackled by individual communities. Ongoing development outside the urban core typically creates many separate communities that are legally independent of each other.

As seen in table 5.2, the result is political fragmentation. For example, the New York City metropolitan area extends over twenty-seven counties in three states and contains more than 21 million people. Taking note of some other urban areas, the Chicago metropolitan area contains more than 1,100 government units, the Philadelphia metropolitan area has close to 900 government units, and Pittsburgh has more than 800 units.

According to Anthony Downs, the existence of many independent suburbs establishes a hierarchy of prestige that reinforces separation:

> Each [jurisdiction] presents a different combination of local tax rates and policies, public services, housing prices, and residents' socioeconomic levels and ethnicity. Tax rates and public services are controlled by local government; housing prices and the socioeconomic status and ethnicity of residents by individual decisions about where to live. Those decisions also affect local government policies because suburban governments are very responsive to residents' desires. In turn, local government policies affect households' decisions about where to live. These interdependent forces encourage the differentiation of suburbs.[26]

As a consequence, suburban income groups tend to segregate themselves according to income levels: High-income households tend to cluster in

Table 5.2.   **Political Fragmentation in the Largest Metropolitan Areas**

Metropolitan Area	Counties	Municipalities & Townships	Total Local Governments	Local Governments per 100,000 Residents
Pittsburgh	6	412	418	17.7
Minneapolis–St. Paul	13	331	344	12.3
St. Louis	12	300	312	12.2
Cincinnati	13	222	235	12.2
Kansas City	11	171	182	10.6
Cleveland	8	259	267	9.2
Philadelphia	14	428	442	7.4
Milwaukee	5	108	113	6.9
Chicago	13	554	567	6.6
Detroit	10	325	335	6.2
Boston	14	282	296	5.1
Dallas	12	184	196	4.2
Portland, Ore.	8	79	87	4.1
New York	27	729	756	3.8
Atlanta	20	107	127	3.5
Denver	7	67	74	3.2
Houston	8	115	123	2.8
Seattle	6	88	94	2.8
Tampa	4	35	39	1.8
San Francisco	10	104	114	1.7
Miami	2	55	57	1.6
Phoenix	2	32	34	1.2
Los Angeles	5	177	183	1.2
San Diego	1	18	19	0.7
Washington-Baltimore	33	125	158	2.2

Source: Myron Orfield, *American Metropolitics: The New Suburban Reality* (Washington, D.C.: Brookings Institution Press, 2003).

high-prestige locations, middle-income families in middling-prestige locations, and low-income households are usually relegated to low-prestige
locations because they cannot afford other options. Thus, housing and living conditions in the suburbs tend to reflect the socioeconomic status of
residents. Moreover, suburban jurisdictions use their powers of zoning
and building codes to keep social distance from the central city as well as
from potential residents of different social and economic backgrounds.

A fairly new trend in suburbia is the introduction of gated communities, which promise safety and security (e.g., guards at the entrance gates)
as well as architectural harmony and the imposition of rules on residents
and visitors. At last count, an estimated eight million Americans lived in
gated communities, mostly around the Los Angeles, Phoenix, Chicago,

Houston, New York, and Miami metropolitan areas. Like most shopping centers and malls, gated communities operate under private government. They are typically run by self-governing homeowner associations, a feature they share with private street subdivisions everywhere. Elected boards oversee common property—such as streets, sidewalks, and common facilities, including the gates—and each home is bound to covenants, conditions, and restrictions as part of its deed. In building such communities, developers appeal

> to the homeowners' natural instinct to preserve and protect their investments from the unwanted interference of local government—such as zoning changes to allow multiple dwellings, commercial facilities, or group homes. Suburban neighborhoods have been transformed into collectively owned property in part to circumvent government regulation and social responsibility.[27]

In a study by the Brookings Institution titled "Pulling Apart," it was found that the gap between the richest and poorest places (cities and suburbs) in major metropolitan areas increased rapidly during the 1980s and more slowly in the 1990s, although patterns of inequality varied widely across the country.[28] Generally, the income gaps were largest in metropolitan areas in the Southwest in such areas as Phoenix, Los Angeles, and Houston. In the northeastern part of the country, such metropolitan areas as Buffalo, Rochester, and Hartford showed less of an income gap between places.

In his book *Cities without Suburbs*, David Rusk contends that metropolitan areas in which central cities have been able to expand through annexation or consolidation with counties have experienced more favorable social and economic results than those in which annexation or consolidation is limited.[29] Such cities were less segregated by race and class and more fiscally sound. In addition, recent literature shows that central cities and suburbs are economically interdependent and that separation is not necessarily to the advantage of suburban communities. In the Ledebur and Barnes study "All in It Together: Cities, Suburbs and Local Economic Regions," it was found that median household incomes of central cities and suburbs moved up and down together; one did not prosper without the other.[30] Another study by H. V. Savitch found that suburban towns that encircle a healthy city are far more likely to succeed economically than those surrounding troubled cities.[31] In light of Third Wave technology, furthermore, a number of scholars contend that metropolitan regions are increasingly the new unit of global competitiveness.[32]

The question to be asked, then, is whether regions can reconfigure their governance to exploit changes taking place in the new economy. Unfortunately, as cities such as Boston, Chicago, Detroit, and Pittsburgh became encircled by incorporated areas, they long ago lost the opportunity to expand their boundaries. At the present time, fewer than a dozen of the 370 metropolitan areas in the country have any form of regional governance. Instead, there tends to be growing reliance on less comprehensive approaches such as interlocal service contracts and joint power agreements in which two or more jurisdictions agree to plan, finance, and deliver a service. This typically takes the form of a special district government to provide a single service, such as sewer, water, recreation, housing, or mosquito-control services.

Because they are voluntary, a politically inoffensive way of doing metropolitan-wide planning is through councils of government (COGs). Each local government is represented in the COG by its own elected officials. Members of COGs meet to discuss problems, exchange information, and make policy proposals for metropolitan development. In the 1980s, there were as many as 660 such agencies that performed an important role of reviewing federal programs being implemented on the local level; however, the numbers declined in the 1990s when federal support was withdrawn.

Though numerous scholars contend that the real city is the metropolitan region, typically consisting of a central city and the surrounding suburbs, governance of metropolitan areas continues to lag, with the likely result that political fragmentation and resulting diseconomies will persist. For the record, we should note that some scholars contend that fragmentation is good, because it encourages competition among jurisdictions. They explain that as in business, competition leads to innovation and provides incentives to produce government services more economically.[33] However, critics reply that many people are not free to act as consumers and do not move when dissatisfied with services, since family, finances, and jobs greatly reduce mobility. Low-income households and the very poor are especially disadvantaged.

The Prognosis for Sense of Community

In referring to "sense of community," the interest here is community rooted in social relationships involving a sense of shared identity and interdependence. This implies that people who share spatial boundaries may not constitute a community where a common identity does not exist. The basic question being posed is: How has the Third Wave era affected sense of community?

In his book *Bowling Alone: The Collapse and Revival of American Community*, Robert D. Putnam reports on how Americans have become increasingly disconnected from family, friends, neighbors, and democratic institutions. Like other social scientists, Putnam contends that a key feature that underlies the well-being of any society or community is *social capi-*

tal.[34] "Whereas physical capital refers to physical objects and human capital refers to properties of individuals, social capital refers to connections among individuals—social networks and the norms of reciprocity and trustworthiness that arise from them."[35] Putnam warns that social capital has plummeted in the nation, thereby impoverishing the community.

Drawing on nearly 500,000 interviews, Putnam shows that people are less willing to sign petitions, belong to fewer organizations that meet, are less knowing of their neighbors, meet with friends less often, are less inclined to bowl in leagues (they bowl alone), and even socialize with their families less often. He analyzes a number of factors that account for this trend, including pressures of time and money on two-career families and suburban commuting. But what weighs in more heavily is what Putnam calls "the technological transformation of leisure." He refers to data that show the growth in time spent watching television. This has dwarfed all other changes in the way Americans pass their days and nights. And while electronic media, including the Internet and e-mail, allow a wide-ranging exploration of information and choice, it tends to be socially isolating and destructive of civic engagement.

It appears, furthermore, that generational change is at least equally important in accounting for the decline of civic engagement. Viewing the 1990s, Putnam shows that almost all forms of civic engagement—from union membership to church attendance to public meeting attendance—have continued to plummet among young people who are in their twenties. In some contrast to their elders, they tend to be much more visually oriented—that is, "perpetual surfers, multitaskers, interactive media specialists."[36]

A counterargument to Putnam, made by Joseph C. Licklider and Robert W. Taylor, projects that "life will be happier for the on-line individual because the people with whom one interacts most strongly will be selected more by commonality of interests and goals than by accidents of proximity."[37] Apart from the question of happiness, Licklider and Taylor are undoubtedly right in contending that online communication facilitates the creation of groups with shared interests. As noted in chapter 3, Internet communications fill a range of needs for individuals. For some, the exchange of information and the sharing of interests is an end in itself. For others, the sharing of information serves important personal or professional goals such as getting legal or medical advice or doing research.

In recent years, new types of electronic message boards called *blogs* have come into existence (see box 5.2). Today, there are blogs on almost any subject, ranging from sex blogs to drug blogs to teenage blogs. There are also news blogs and commentary blogs dealing with local, national, and international issues.

Box 5.2. The Blogging Revolution

A *blog* is a website for which an individual or a group produces text, photographs, video or audio files, and/or links, usually on a daily basis. The term is a shortened form of "Web log." Creating a blog or adding an article to an existing blog is called "blogging." Individual articles on a blog are called "blog posts," "posts," or "entries." The person who posts these entries is a "blogger."

In the beginning, around 1994, what we now call blogging was little more than the inspired writing of online diaries. But as more users have logged on, it is changing the media world and could eventually foment a revolution, replacing more traditional methods of communicating. By the end of 2004, blogs had established themselves as a key feature of online culture. Two surveys by the Pew Internet and American Life Project came up with the following profile for blogging: 8 million American adults said they had created blogs; blog readership jumped 58 percent in 2004 and represented 27 percent of Internet users; 5 percent of Internet users said they get the news and other information delivered from blogs as it is posted online; and 12 percent of Internet users have posted comments or other material on blogs. It is of interest to note, furthermore, that 62 percent of Internet users reported that they did not know what a blog is.

Source: "The State of Blogging," memorandum, Pew Internet and American Life Project, January 2005, www.pewinternet.org/PIP_blogging_data.pdf.

But are these shared activities "communities"? William A. Galston expresses doubt. He argues that a typical feature of online groups is weak control over their members: Anyone can enter and anyone can leave at any time. Because exiting is easy, members are inclined to break away and start new groups when dissatisfied with some aspects of their group. According to Galston, this has certain negative consequences.

> In a diverse democratic society, politics requires the ability to deliberate, and compromise with individuals unlike oneself. When we find ourselves living cheek by jowl with neighbors with whom we differ but whose propinquity we cannot easily escape, we have powerful incentives to develop modes of accommodation. On the other hand, the ready availability of exit tends to produce internally homogeneous groups that may not communicate with other groups and lack incentives to develop shared understanding across their differences.[38]

Another distinction to be considered is the role of visual and tonal cues in establishing ties of depth and significance between individuals. An underlying assumption is that in human interaction, persons naturally rely on a range of nonverbal evidence to assess the motivations and sincerity of others. Electronic text communication does not provide for this. Furthermore, it can be argued that because the absence of visual and tonal cues makes it difficult to see the human effects that words can inflict, the Internet does not effectively restrain impulsive verbal behavior.

Looking to the future, electronic technology will continue to have a major impact on how persons interact with others. The challenge that lies ahead will be to foster new, more innovative ways of using such technology to facilitate community engagement. For example, how can the Internet be designed to allow real face-to-face communities and not merely to displace them with virtual community? We also see where computer-mediated communication in the form of blogs is able to open opportunities for new forms of community building such as facilitating citywide citizen debates on important issues, addressing the special needs of the elderly and the sick, or something as simple as notifying residents of upcoming civic events.

The Prognosis for Core Cities and Older Metros

Another critical question to be considered is: How well will core cities and older metropolitan areas adapt to the postmodern era? Such prominent futurists as John Naisbitt and Alvin Toffler have predicted that core cities are doomed and that new telecommunications have made human interaction based on physical proximity superfluous.[39] Taking exception to that prognosis, Anthony Downs contends that the future of core city economies will depend on the degree to which they maintain and strengthen those activities in which they possess a comparative advantage.

Where *do* older core cities possess a comparative advantage? Consider the following factors:

- Though managerial and professional offices continue to disperse into the hinterland of metropolitan areas, many remain concentrated in central cities because these locations facilitate face-to-face communication that is so important in high-level decision making. This assumes that technology's ability to substitute for face-to-face contact in high-level decision making will remain limited.
- Major segments of the population continue to be attracted to lifestyles that draw sustenance from concentrated urban living—walking cities, lively street life, and easily accessible cultural offerings.
- Environmental factors, such as demand for clean air and water, encourage more concentrated patterns of development that reduce auto travel and preserve open space. This creates growing pressure for development at the core.

- Because of differential productivity growth and an increasing global division of labor, complex activities continue to parallel or outpace the growth of routine activities. When tasks are complicated and instructions can be easily misunderstood, close physical contact is important in facilitating communication.
- Core cities provide a more dynamic environment for innovation. This includes ready access to specialized skills, detailed market knowledge, and support services needed for the development of new products and services.
- Urbanized economies derived from diverse job markets, labor markets, supplier markets, and air transport tend to support location in large metropolitan areas.

However, the ability of core areas to adapt to postindustrial change is limited by factors that pull in the opposite direction.

- Business and industry will continue to locate in lower-cost, higher-amenity areas, causing additional job loss in urban core areas that evidence high cost of living and low quality of life.
- Advances in technology are likely to increase telecommuting and the ability of firms to locate in the far reaches of urban regions.
- Advances in telephony are likely to serve as acceptable substitutes for face-to-face activities.
- Social problems such as crime, poverty, and poor schooling in core cities encourage the continuing exodus of middle-class residents.
- Many core cities continue to struggle with declining revenues and simultaneously rising demand for services.
- Technology is likely to continue increasing the mismatch of skills between higher-end jobs in cities and the lower-skilled labor force that lives there.

But while technology is generally viewed as posing constraints on the urban core, technology can also be designed to strengthen the urban core. Among those innovations already in use are intelligent transportation systems, video surveillance for public safety, electronic delivery of services, electronic data retrieval, geophysical positioning systems, and the use of telecommunications to promote citizen participation.

World history is replete with stories of the rise and fall of powerful city-states that dominated vast global regions for long periods of time. Athens led its allies into the victorious wars against Persia only to inherit an empire that later fragmented and was replaced by the might of Alexandria. Subsequently, the city-state of Rome controlled a vast area stretching from the Sahara in the south and the Euphrates in the east to Britain in the west. Rome's decline made way for the Goths, Vandals, and other invaders from whom modern European states emerged. Thus, history teaches that even the most powerful cities must be able to adapt to change if they are to remain viable. Those that cannot must pay the price.

In viewing the history of American cities from preindustrialization to industrialization to postindustrialization, we see that some have prospered and grown in making the transition to the new era, while others have languished. Such a transition depends heavily on such factors as location, resources, and demographics. Much also depends on the quality and organization of public–private leadership in these cities. This, then, is one of the fundamental tensions that will be carried into the postmodern era.

For most of history, humans have shaped civilization commensurate with the level of their material technology. In some cases, like that of ancient Athens, civilization far outstripped technology. New technologies will inevitably proliferate in the years to come, but we need to remember that the measure of a civilization is not the tools it owns, but the use it makes of them.

NOTES

1. Robert E. Park, Ernest W. Burgess, and Roderick D. McKenzie, *The City* (Chicago: University of Chicago Press, 1925).

2. Amos H. Hawley, *Urban Society: An Ecological Approach* (New York: Ronald Press, 1971), 99.

3. U.S. Federal Housing Administration, *The Structure and Growth of Residential Neighborhoods in American Cities* (Washington, D.C.: GPO, 1939).

4. P. M. Rees, "Problems of Classifying Sub-areas within Cities," in *City Classification Handbook: Methods and Applications*, ed. Brian J. L. Berry (New York: Wiley-Interscience, 1972), 265–330; Eshref Shevky and Wendell Bell, *Social Area Analysis: Theory, Illustrative Application, and Computational Procedures* (Stanford, Calif.: Stanford University Press, 1955).

5. Brian J. L. Berry and John D. Kasarda, *Contemporary Urban Ecology* (New York: Macmillan, 1977), 7.

6. Mark Gottdiener and Joe R. Feagin, "The Paradigm Shift in Urban Sociology," *Urban Affairs Quarterly* 24, no. 2 (December 1988): 163–87.

7. John R. Logan and Harvey L. Molotch, *Urban Fortunes: The Political Economy of Place* (Berkeley: University of California Press, 1987), 52.

8. Clarence N. Stone, *Regime Politics: Governing Atlanta, 1946–1988* (Lawrence: University Press of Kansas, 1989); Richard Edward DeLeon, *Left Coast City: Progressive Politics in San Francisco, 1975–1991* (Lawrence: University Press of Kansas, 1992).

9. Stone, *Regime Politics*, 290.

10. Neal R. Peirce, with Curtis W. Johnson and John Stuart Hall, *Citistates: How Urban America Can Prosper in a Competitive World* (Washington, D.C.: Seven Locks Press, 1993).

11. Saskia Sassen, *Cities in a World Economy*, 3rd ed. (Thousand Oaks, Calif.: Pine Forge Press, 2006); Saskia Sassen, *The Global City: New York, London, Tokyo*, 2nd ed. (Princeton, N.J.: Princeton University Press, 2001).

12. Dennis A. Rondinelli, James Johnson, and John D. Karsada, "The Changing Forces of Urban Economic Development: Globalization and City Competitiveness in the 21st Century," *Citiscape* 3 (1998): 71–106.

13. Peirce, *Citistates*, 37, 38.

14. Heywood T. Sanders, "Convention Center Follies," *Public Interest* 132 (Summer 1998): 58–72.

15. Sanders, "Convention Center Follies," 61.

16. Mark F. Bernstein, "Sports Stadium Boondoggle," *The Public Interest* 132 (Summer 1998): 45–57.

17. See Jane Jacobs, *The Death and Life of Great American Cities* (New York: Vintage, 1961).

18. See, for example, André Sorensen, Peter J. Marcotullio, and Jill Grant, eds., *Towards Sustainable Cities: East Asian, North American, and European Perspectives on Managing Urban Regions* (Burlington, Vt.: Ashgate, 2004).

19. Lawrence J. R. Herson and John M. Bolland, *The Urban Web: Politics, Policy, and Theory*, 2nd ed. (Chicago: Nelson-Hall, 1998), 234.

20. Roger R. Stough and Jean Paelinck, "Substitution and Complementary Effects of Information on Regional Travel and Location Behavior," paper presented at the Regional Science International Association World Congress, Tokyo, 1–5 May 1996.

21. Robert E. Lang, *Edgeless Cities: Exploring the Elusive Metropolis* (Washington, D.C.: Brookings Institution Press, 2003).

22. Lang, *Edgeless Cities*, 55.

23. U.S. Department of Housing and Urban Development, *The State of the Cities 2000* (Washington, D.C.: HUD, 2000), 40, 41.

24. See Mike Davis, *City of Quartz: Excavating the Future in Los Angeles* (New York: Verso, 1990); Mike Davis, *Ecology of Fear: Los Angeles and the Imagination of Disaster* (New York: Metropolitan Books, 1998).

25. Peter Whoriskey, "Investing in Sprawl: The Limits of Smart Growth," *Washington Post*, 10 August 2004.

26. Anthony Downs, *New Visions for Metropolitan America* (Washington, D.C.: Brookings Institution, 1994), 21.

27. Edward J. Blakely and Mary Gail Snyder, *Fortress America: Gated Communities in the United States* (Washington, D.C.: Brookings Institution Press, 1998), 159.

28. Todd Swanstrom, Colleen Casey, Robert Flack, and Peter Dreier, "Pulling Apart: Economic Segregation among Suburbs and Central Cities in Major Metropolitan Areas" (Washington, D.C.: Brookings Institution, 2004), available at www.brookings.edu/metro/pubs/20041018_econsegregation.htm.

29. David Rusk, *Cities without Suburbs: A Census 2000 Update*, 3rd ed. (Washington, D.C.: Woodrow Wilson Center Press, 2003).

30. Henry G. Cisneros, ed., *Interwoven Destinies: Cities and the Nation* (New York: Norton, 1993), 23–24.

31. H. V. Savitch, "Ties That Bind: Central Cities, Suburbs and the New Metropolitan Region," paper presented at the annual meeting of the American Political Science Association, Chicago, 3–6 September 1992.

32. See Peirce, *Citistates*; Sassen, *Cities in a World Economy*; and Sassen, *The Global City*.

33. Vincent Ostrom, Charles Tiebout, and Robert Warren, "The Organization of Government in Metropolitan Areas," *American Political Science Review* (December 1961): 831–42.

34. The first theorist to use the term *social capital* was Lyda Judson Hanifan, in "The Rural School Community Center," *Annals of the American Academy of Political and Social Science* 67 (1916): 130–38. See also Jacobs, *Death and Life*; Ronald S. Burt, "The Contingent Value of Social Capital," *Administrative Science Quarterly* 42 (1997): 339–65.

35. Robert D. Putnam, *Bowling Alone: The Collapse and Revival of American Community* (New York: Simon & Schuster, 2000), 19.

36. Putnam, *Bowling Alone*, 259.

37. Quoted in Steven G. Jones, ed., *Virtual Culture: Identity and Communication in Cybersociety* (Thousand Oaks, Calif.: Sage, 1997).

38. William A. Galston, "Does the Internet Strengthen Community?" *Philosophy and Public Policy* 19, no. 4 (Fall 1999): 5.

39. John Naisbitt, *Megatrends* (New York: Warner Books, 1982); John Naisbitt, *Global Paradox: The Bigger the World Economy, the More Powerful Its Smallest Players* (New York: Morrow, 1994); Alvin Toffler and Heidi Toffler, *Creating a New Civilization: The Politics of the Third Wave* (Atlanta: Turner Publishing, 1994).

Bibliography

Abrahamson, Mark. *Global Cities*. New York: Oxford University Press, 2004.

Abu-Lughod, Janet L. *New York, Chicago, Los Angeles: America's Global Cities*. Minneapolis: University of Minnesota Press, 1999.

Alavi, Maryam. "Dick Tracy's Office—Business Applications of Wireless Technologies." In *The Emerging World of Wireless Communications*, ed. C. M. Firestone, 133–42. Queenstown, Md.: Aspen Institute and Institute for Information Studies, 1996.

Allen, Frederick Lewis. *The Big Change: America Transforms Itself, 1900–1950*. New York: Harper, 1952.

Bell, Daniel. *The Coming of Post-Industrial Society: A Venture in Social Forecasting*. New York: Basic Books, 1976.

Berry, Brian J. L. *The Human Consequences of Urbanization: Divergent Paths in the Urban Experience of the Twentieth Century*. New York: St. Martin's, 1973.

Berry, Brian J. L., and John D. Kasarda. *Contemporary Urban Ecology*. New York: Macmillan, 1977.

Blakely, Edward J., and Mary Gail Snyder. *Fortress America: Gated Communities in the United States*. Washington, D.C.: Brookings Institution Press, 1998.

Bridenbaugh, Carl. *Cities in the Wilderness: The First Century of Urban Life in America, 1625–1742*. New York: Ronald Press Co., 1938.

Brookings Institution. Center for Urban and Metropolitan Policy. *Growing the Middle Class: Connecting All Miami-Dade County Residents to Economic Opportunity*. Washington, D.C.: Brookings Institution, 2004. Available at www.brookings.edu/metro/publications/20040607_miami.htm.

———. *Los Angeles in Focus: A Profile from Census 2000*. Washington, D.C.: Brookings Institution, 2003. Available at www.brookings.edu/es/urban/livingcities/losangeles.htm.

———. *Racial Change in the Nation's Largest Cities: Evidence from the 2000 Census.* Washington, D.C.: Brookings Institution, 2001. Available at www.brookings.edu/es/urban/census/citygrowth.htm.

Burt, Ronald S. "The Contingent Value of Social Capital." *Administrative Science Quarterly* 42 (1997): 339–65.

Cairncross, Frances. *The Death of Distance: How the Communications Revolution Will Change Our Lives.* Boston: Harvard Business School Press, 1997.

Castells, Manuel. *The Informational City: Information Technology, Economic Restructuring, and the Urban-Regional Process.* Cambridge, Mass.: Blackwell, 1989.

Castells, Manuel, and Peter Hall. *Technopoles of the World: The Making of Twenty-First-Century Industrial Complexes.* New York: Routledge, 1994.

Christopherson, S., and M. Storper. "The Effects of Flexible Specialization on Industrial Politics and the Labor Market: The Motion Picture Industry." *Industrial and Labor Relations Review* (1989): 331–47.

Cisneros, Henry G., ed. *Interwoven Destinies: Cities and the Nation.* New York: Norton, 1993.

Davis, Mike. *City of Quartz: Excavating the Future in Los Angeles.* New York: Verso, 1990.

———. *Ecology of Fear: Los Angeles and the Imagination of Disaster.* New York: Metropolitan Books, 1998.

Dear, Michael, and Steven Flusty. "The Iron Lotus: Los Angeles and Postmodern Urbanism." In *Globalization and the Changing U.S. City*, ed. David Wilson, 151–63. Thousand Oaks, Calif.: Sage, 1997.

DeLeon, Richard Edward. *Left Coast City: Progressive Politics in San Francisco, 1975–1991.* Lawrence: University Press of Kansas, 1992.

Downs, Anthony. *New Visions for Metropolitan America.* Washington, D.C.: Brookings Institution, 1994.

Drezner, Daniel. "The Outsourcing Bogeyman." *Foreign Affairs* 83, no. 3 (May–June 2004): 24.

Dutton, William H., Jay G. Blumler, and Kenneth L. Kraemer, eds. *Wired Cities: Shaping the Future of Communications.* Boston: G. K. Hall, 1988.

Fathy, Tarik A. *Telecity: Information Technology and Its Impact on City Form.* New York: Praeger, 1991.

Findlay, John M. *Magic Lands: Western Cityscapes and American Culture after 1940.* Berkeley: University of California Press, 1992.

Fishman, Robert. "America's New City: Megalopolis Unbound." *Wilson Quarterly* (Winter 1990): 38.

Florida, Richard. *The Rise of the Creative Class: And How It's Transforming Work, Leisure, Community and Everyday Life.* New York: Basic Books, 2004.

Freidman, Thomas L. *The World Is Flat: A Brief History of the Twenty-First Century.* New York: Farrar, Straus and Giroux, 2005.

Gabriel, Stuart A. "Remaking the Los Angeles Economy." In *Rethinking Los Angeles*, ed. Michael Dear, H. Eric Schockman, and Greg Hise, 22–33. Thousand Oaks, Calif.: Sage, 1996.

Galston, William A. "Does the Internet Strengthen Community?" *Philosophy and Public Policy* 19, no. 4 (Fall 1999): 5.

Garreau, Joel. *Edge City: Life on the New Frontier.* New York: Doubleday, 1991.

Gates, Bill, with Nathan Myhrvold and Peter Rinearson. *The Road Ahead*. New York: Viking, 1995.

Gay, Martin, and Kathlyn Gay. *The Information Superhighway*. New York: Twenty-first Century Books, 1996.

Goldfield, David R., and Blaine A. Brownell. *Urban America: A History*. 2nd ed. Boston: Houghton Mifflin, 1990.

Gottdiener, Mark, and Joe R. Feagin. "The Paradigm Shift in Urban Sociology." *Urban Affairs Quarterly* 24, no. 2 (December 1988): 163–87.

Gottman, Jean. *Megalopolis: The Urbanized Northeastern Seaboard of the United States*. New York: Twentieth Century Fund, 1961.

Gurwitt, Rod. "Black Mayors and the Board Room." *Governing* 7 (April 1996): 42.

Hafner, Katie, and Matthew Lyon. *Where Wizards Stay Up Late: The Origins of the Internet*. New York: Simon & Schuster, 1996.

Hanifan, Lyda Judson. "The Rural School Community Center." *Annals of the American Academy of Political and Social Science* 67 (1916): 130–38.

Hannigan, John. *Fantasy City: Pleasure and Profit in the Postmodern Metropolis*. New York: Routledge, 1998.

Hartshorn, Truman A., and Peter O. Muller, "Suburban Downtowns and the Transformation of Metropolitan Atlanta's Business Landscape." *Urban Geography* 10 (1989): 375–95.

Hatfield, David N. *The Technological Basis for Wireless Communications*. Queenstown, Australia: Institute for Information Studies, 1996.

Hawley, Amos H. *Urban Society: An Ecological Approach*. New York: Ronald Press, 1971.

Herson, Lawrence J. R., and John M. Bolland. *The Urban Web: Politics, Policy, and Theory*. 2nd ed. Chicago: Nelson-Hall, 1998.

Hillman, Judy. *Telelifestyles and the Flexicity: A European Study—The Impact of the Electronic Home*. Dublin: European Foundation for the Improvement of Living and Working Conditions, 1993.

Huxtable, Ada Louise. *The Unreal America: Architecture and Illusion*. New York: New Press, 1997.

Jackson, Kenneth T. *Crabgrass Frontier: The Suburbanization of the United States*. New York: Oxford University Press, 1985.

Jacobs, Jane. *The Death and Life of Great American Cities*. New York: Vintage, 1961.

Jones, Steven G., ed. *Virtual Culture: Identity and Communication in Cybersociety*. Thousand Oaks, Calif.: Sage, 1997.

Kaplan, J. "Rooting for a Logo." In *The City and the World: New York's Global Future*, ed. Margaret E. Crahan and Alberto Vourvoulais-Bush, 159–70. New York: Council on Foreign Relations, 1997.

Kaplan, Robert K. "Travels into America's Future." *Atlantic Monthly* 282 (July 1998): 56–58.

Katz, Bruce. "Smart Growth: The Future of the American Metropolis." Case Paper 58. London: Center for Analysis of Social Exclusion, London School of Economics, 2002.

Kendall, Edward Augustus. *Travels through the Northern Parts of the United States, in the Years 1807 and 1808*. 3 vols. New York: I. Riley, 1809.

Kenney, Martin, ed. *Understanding Silicon Valley: The Anatomy of an Entrepreneurial Region*. Stanford, Calif.: Stanford University Press, 2000.

Kurshan, B., and C. Lenk, "The Technology of Learning." In *Crossroads on the Information Highway: Convergence and Diversity in Communications Technologies*, 120–31. Annual review of the Institute for Information Studies, Aspen Institute, and Northern Telecom. Queenstown, Md.: Institute for Information Studies, 1995.

Lang, Robert E. *Edgeless Cities: Exploring the Elusive Metropolis*. Washington, D.C.: Brookings Institution Press, 2003.

Le Corbusier. *The City of To-morrow and Its Planning*. Cambridge, Mass.: MIT Press, 1971.

Lemow, Penelope. "Welcome to Eldertown." *Governing* 10 (October 1996): 23.

Levinson, Paul. *Cellphone: The Story of the World's Most Mobile Medium and How It Has Transformed Everything*. New York: Palgrave Macmillan, 2004.

Logan, John R., and Harvey L. Molotch. *Urban Fortunes: The Political Economy of Place*. Berkeley: University of California Press, 1987.

Longino, Charles F. Jr. *Retirement Migration in America*. Houston: Vacation Publications, 1995.

MacManus, Susan A., with Patricia A. Turner. *Young v. Old: Generational Combat in the 21st Century*. Boulder, Colo.: Westview Press, 1996.

McLuhan, Marshall. *Understanding Media: The Extensions of Man*. New York: McGraw-Hill, 1965.

McLuhan, Marshall, and Bruce R. Powers. *The Global Village: Transformations in World Life and Media in the 21st Century*. New York: Oxford University Press, 1989.

Meyrowitz, Joshua. *No Sense of Place: The Impact of Electronic Media on Social Behavior*. New York: Oxford University Press, 1985.

Mitchell, William J. *City of Bits: Space, Place, and the Infobahn*. Cambridge, Mass.: MIT Press, 1995.

———. *E-topia*. Cambridge, Mass.: MIT Press, 1999.

Mollenkopf, John Hull, and Manuel Castells, eds. *Dual City: Restructuring New York*. New York: Russell Sage Foundation, 1991.

Montiel, Lisa M., Richard P. Nathan, and David J. Wright. *An Update on Urban Hardship*. Albany, N.Y.: Nelson A. Rockefeller Institute of Government, 2004.

Moore, J. F. "Convergence and the Development of Business Ecosystems." In *Crossroads on the Information Highway: Convergence and Diversity in Communications Technologies*, 132–42. Annual review of the Institute for Information Studies, Aspen Institute, and Northern Telecom. Queenstown, Md.: Institute for Information Studies, 1995.

Mumford, Lewis. *The Culture of Cities*. New York: Harcourt, Brace, 1938.

———. *The Highway and the City*. New York: Harcourt, Brace & World, 1963.

Naisbitt, John. *Global Paradox: The Bigger the World Economy, the More Powerful Its Smallest Players*. New York: Morrow, 1994.

———. *Megatrends*. New York: Warner Books, 1982.

Nye, David E. *American Technological Sublime*. Cambridge, Mass.: MIT Press, 1994.

Ohmae, Kenichi. *The End of the Nation State: The Rise of Regional Economies*. New York: Free Press, 1995.

Orfield, Myron. *American Metropolitics: The New Suburban Reality.* Washington, D.C.: Brookings Institution Press, 2003.

Ostrom, Vincent, Charles Tiebout, and Robert Warren. "The Organization of Government in Metropolitan Areas." *American Political Science Review* (December 1961): 831–42.

Park, Robert E., Ernest W. Burgess, and Roderick D. McKenzie. *The City.* Chicago: University of Chicago Press, 1925.

Peirce, Neal R., with Curtis W. Johnson and John Stuart Hall. *Citistates: How Urban America Can Prosper in a Competitive World.* Washington, D.C.: Seven Locks Press, 1993.

Pratt, J. H. "Home Teleworking: A Study of Its Pioneers." *Technological Forecasting and Social Change* 25 (1984): 1–14.

Pred, Allan R. *The Spatial Dynamics of U.S. Urban Industrial Growth, 1800–1914.* Cambridge, Mass.: MIT Press, 1966.

Putnam, Robert D. *Bowling Alone: The Collapse and Revival of American Community.* New York: Simon & Schuster, 2000.

Rees, P. M. "Problems of Classifying Sub-areas within Cities." In *City Classification Handbook: Methods and Applications,* ed. Brian J. L. Berry, 265–330. New York: Wiley-Interscience, 1972.

Reich, Robert B. *The Work of Nations: Preparing Ourselves for Twenty-First-Century Capitalism.* New York: Vintage Books, 1992.

Rogers, Everett M. *Diffusion of Innovations.* 3rd ed. New York: Free Press, 1983.

Rondinelli, Dennis A., James Johnson, and John D. Karsada. "The Changing Forces of Urban Economic Development: Globalization and City Competitiveness in the 21st Century." *Citiscape* 3 (1998): 71–106.

Rusk, David. *Cities without Suburbs: A Census 2000 Update.* 3rd ed. Washington, D.C.: Woodrow Wilson Center Press, 2003.

Sanders, Heywood T. "Convention Center Follies." *Public Interest* 132 (Summer 1998): 58–72.

Sassen, Saskia. *Cities in a World Economy.* 3rd ed. Thousand Oaks, Calif.: Pine Forge Press, 2006.

———. *The Global City: New York, London, Tokyo.* 2nd ed. Princeton, N.J.: Princeton University Press, 2001.

Schlesinger, Arthur Meier. *The Rise of the City, 1878–1898.* New York: Macmillan, 1933.

Schultz, Stanley K., and Clay McShane. "To Engineer the Metropolis: Sewers, Sanitation, and City Planning in Late Nineteenth-Century America." *Journal of American History* 65 (September 1978).

Schwartz, David. *Suburban Xanadu: The Casino Resort on the Las Vegas Strip and Beyond.* New York: Routledge, 2003.

Scott, A. J. "Flexible Production Systems: The Rise of New Industrial Spaces in North America and Western Europe." *International Journal of Urban Regional Research* 2 (1988): 171–85.

———. *Technopolis: High-Technology Industry and Regional Development in Southern California.* Berkeley: University of California Press, 1993.

Scott, Mellier. *American City Planning.* Berkeley: University of California Press, 1969.

Shevky, Eshref, and Wendell Bell. *Social Area Analysis: Theory, Illustrative Application, and Computational Procedures.* Stanford, Calif.: Stanford University Press, 1955.

Simon, Bryant. *Boardwalk of Dreams: Atlantic City and the Fate of Urban America.* New York: Oxford University Press, 2004.

Soja, Edward W. *Postmodern Geographies: The Reassertion of Space in Critical Social Theory.* London: Verso, 1989.

———. "Poles Apart: Urban Restructuring in New York and Los Angeles." In *Dual City: Restructuring New York*, ed. John Hull Mollenkopf and Manuel Castells, 359–76. New York: Russell Sage Foundation, 1991.

Sorensen, André, Peter J. Marcotullio, and Jill Grant, eds. *Towards Sustainable Cities: East Asian, North American, and European Perspectives on Managing Urban Regions.* Burlington, Vt.: Ashgate, 2004.

Squires, Gregory D., ed. *Urban Sprawl.* Washington, D.C.: Urban Institute Press, 2002.

Sternlieb, George, and James W. Hughes. *The Atlantic City Gamble.* Cambridge, Mass.: Harvard University Press, 1983.

Stewart, Doug. "Times Square Reborn." *Smithsonian* 28, no. 11 (February 1998): 36–37.

Stoll, Michael A. "Job Sprawl and the Spatial Mismatch between Blacks and Jobs." Washington, D.C.: Brookings Institution, 2005. Available at www.brookings .edu/metro/pubs/20050214_jobsprawl.pdf.

Stone, Clarence N. *Regime Politics: Governing Atlanta, 1946–1988.* Lawrence: University Press of Kansas, 1989.

Swanstrom, Todd, Colleen Casey, Robert Flack, and Peter Dreier. "Pulling Apart: Economic Segregation among Suburbs and Central Cities in Major Metropolitan Areas." Washington, D.C.: Brookings Institution, 2004. Available at www .brookings.edu/metro/pubs/20041018_econsegregation.htm.

Taylor, Peter J., and Robert E. Lang. "U.S. Cities in the 'World City Network.'" Washington, D.C.: Brookings Institution, 2005. Available at www.brookings.edu/ metro/pubs/20050222_worldcities.pdf.

Tocqueville, Alexis de. *Democracy in America.* 1835. Reprint. 2 vols. in 1. Ed. J. P. Mayer. Trans. George Lawrence. Garden City, N.Y.: Doubleday, 1969.

Toffler, Alvin. *The Third Wave.* New York: Morrow, 1980.

Toffler, Alvin, and Heidi Toffler. *Creating a New Civilization: The Politics of the Third Wave.* Atlanta: Turner, 1994.

Touraine, Alain. *The Post-Industrial Society; Tomorrow's Social History: Classes, Conflicts and Culture in the Programmed Society.* Trans. Leonard F. X. Mayhew. New York: Random House, 1971.

Trump, Donald. *The Art of the Deal.* New York: Time Warner, 1987.

U.S. Bureau of Labor Statistics. "9/11 and the New York City Economy." *Monthly Labor Review* 127 (June 2004): 1–42.

U.S. Conference of Mayors. *U.S. Metro Economies.* Washington, D.C.: Conference of Mayors, 2001.

U.S. Congress. Office of Technology Assessment. *The Technological Reshaping of Metropolitan America.* Washington, D.C.: GPO, 1995.

U.S. Department of Housing and Urban Development (HUD). *The State of the Cities 2000*. Washington, D.C.: HUD, 2000.

U.S. Federal Housing Administration. *The Structure and Growth of Residential Neighborhoods in American Cities*. Washington, D.C.: GPO, 1939.

Vesperi, Maria D. *City of Green Benches: Growing Old in a New Downtown*. Ithaca, N.Y.: Cornell University Press, 1985.

Vogt, E. E. "The Nature of Work in 2012." In *Crossroads on the Information Highway: Convergence and Diversity in Communications Technologies*, 146–51. Annual review of the Institute for Information Studies, Aspen Institute, and Northern Telecom. Queenstown, Md.: Institute for Information Studies, 1995.

Wachs, Martin, and Margaret Crawford. *The Car and the City: The Automobile, the Built Environment, and Daily Urban Life*. Ann Arbor: University of Michigan Press, 1991.

Wallis, Allan D. "Evolving Structures and Challenges of Metropolitan Regions." *National Civic Review* 83 (Winter–Spring 1994): 40–53.

Warner, Sam Bass Jr. *The Private City: Philadelphia in Three Periods of Its Growth*. Philadelphia: University of Pennsylvania Press, 1968.

Watts, Steven. *The Magic Kingdom: Walt Disney and the American Way of Life*. Boston: Houghton Mifflin, 1997.

Wells, H. G. "The Discovery of the Future." *Nature* 65, no. 1684 (6 February 1902): 328.

White, Morton, and Lucia White. *The Intellectual versus the City, from Thomas Jefferson to Frank Lloyd Wright*. Cambridge, Mass.: Harvard University Press, 1964.

Williams, John Hoyt. *A Great and Shining Road: The Epic Story of the Transcontinental Railroad*. New York: Times Books, 1988.

Index

About the Author

Leonard I. Ruchelman is Eminent Scholar and Professor of Urban Studies and Public Administration at Old Dominion University in Norfolk, Virginia. He has published widely in the general area of urban affairs. His books include *Big City Mayors, Police Politics,* and *The World Trade Center: The Politics and Policies of Skyscraper Development.* He is a member of the Urban Affairs Association and the American Society for Public Administration. He received his BA from Brooklyn College and his PhD from Columbia University. He presently resides with his wife in Virginia Beach, Virginia.